W9-AGB-100

TEACHER'S GUIDE

MATH
in Everyday Life

Third Edition

David E. Newton

WALCH PUBLISHING®

Dedication

For Linda Otto and Jeff Blum,
to help you with the day's aggravations!
All my love and appreciation.

2   3   4   5   6   7   8   9   10

ISBN 0-8251-4283-0

Copyright © 1976, 1991, 2001
J. Weston Walch, Publisher
P.O. Box 658 • Portland, Maine 04104-0658
www.walch.com

Contents

Preface to the Third Edition

Math students throughout the United States and Canada have been helping a typical American family solve its financial problems for 25 years. We are pleased that teachers have found these exercises helpful in introducing students to the kinds of mathematical problems that most people encounter in their everyday lives.

These problems have changed considerably since the first edition of this book appeared in 1976. Today most people make use of hand calculators, computers, and the Internet to think about and solve the mathematical problems they come across in their daily lives. This book has been modified to reflect those changes.

The National Council of Teachers of Mathematics has developed a comprehensive set of standards, *Principles and Standards for School Mathematics,* the *NCTM Standards 2000,* which outlines the skills expected of mathematics students at every grade level from pre-kindergarten through grade 12. Many school systems and most developers of mathematics curriculum turn to this important document for guidelines on grade-level content and for suggestions on how to teach that content.

This revision of *Math in Everyday Life* has been conducted with these *NCTM Standards 2000* in mind.

Number and Operations Standard for Grades 6–8

- Understand numbers, ways of representing numbers, relationships among numbers, and number systems

  - work flexibly with fractions, decimals, and percents to solve problems;

  - understand and use ratios and proportions to represent quantitative relationships;

- Understand meanings of operations and how they related to one another

  - understand the meaning and effects of arithmetic operations with fractions, decimals, and integers;

- Compute fluently and make reasonable estimates

  - select appropriate methods and tools for computing with fractions and decimals from among mental computation, estimation, calculators or computers, and paper and pencil, depending on the situation, and apply the selected methods;

  - develop and analyze algorithms for computing with fractions, decimals, and integers and develop fluency in their use.

Measurement Standard for Grades 6–8

- Understand measurable attributes of objects and the units, systems, and processes of measurement

- understand both metric and customary systems of measurement;

- understand relationships among units and convert from one unit to another within the same system;

- understand, select, and use units of appropriate size and type to measure angles, perimeter, area, surface area, and volume.

Problem Solving Standard for Grades 6–8

Instructional programs from pre-kindergarten through grade 12 should enable all students to—

- build new mathematical knowledge through problem solving;

- solve problems that arise in mathematics and in other contexts;

- apply and adapt a variety of appropriate strategies to solve problems.

Connections Standards for Grades 6–8

Instructional programs from pre-kindergarten through grade 12 should enable all students to—

- recognize and use connections among mathematical ideas;

- recognize and apply mathematics in contexts outside of mathematics.

To the Teacher

This is a *workbook* in consumer mathematics. It is intended to be just that, a collection of exercises on the kinds of mathematical problems that people run into every week of their lives. This is not a textbook in the traditional sense, but instead provides an array and variety of specific, practical problems.

One important objective of this workbook is to make the subject of consumer math relevant to students' lives. To the author, that means using problems and information from students' own lives. In many instances, we suggest that students visit local stores, read local newspapers, collect information from local sources—all in order to do the problems in the workbook. Students use the Internet to widen the search and enrich the experience of using math in everyday life. We think this emphasis will not only improve students' motivation, but will also improve the rate at which they learn.

The problems in the workbook are chosen from the everyday life of a fictitious "average" family, the Van Dusens. We recognize the risk that some of the activities in this book may not be typical of all families. We feel the risk is outweighed by the opportunity of getting to know the Van Dusen family and of feeling that the problems in the workbook are real problems of everyday life.

The workbook is set up to allow teachers to use as much or as little imagination as they wish. Each activity is independent but it is possible to carry over information and data from one chapter to another: The wages reported in Chapter 1, for example, can be used to develop budgets in Chapter 2 and to figure income taxes in Chapter 12. The check-writing exercise can be directed toward paying specific bills found in other sections, and so on. Each teacher can balance the advantages of this realistic approach to learning about home finances with the time available.

You will find multiple copies of some forms. The book contains six copies of Mr. Van Dusen's pay record, for example.

The teacher pages provide answers to nearly every exercise. Sometimes the answer is unique, that is, the only one possible; for example, there is one correct answer for the net pay earned by Mr. and Mrs. Van Dusen and Craig. In other cases, the answers given are only examples of those your students might obtain. Exercises in which students have to visit stores, read advertisements, or make decisions will have many possible correct answers, depending upon prices of the items students find.

Chapter 1

INCOME

Teaching Notes

Many companies now have completely automated payrolls. If possible, demonstrate to students how computer applications can be used to calculate an employee's paycheck almost instantaneously. A parent, local business person, or business teacher might be able to help students use one of these programs to calculate payroll sheets from the student book.

Remind students about the necessity of rounding off monetary calculations with more than two decimal points. An answer such as $89.375, for example, must be rounded off to the nearest penny.

In all calculations, emphasize the importance of estimating and checking answers. Students should understand salaries and wages well enough to know that weekly wages of $1.50 or $15,000.00 are not reasonable, indicating an error in calculation.

Whether calculations are to be done by hand or with calculators depends on your objectives in using this workbook. Hand calculations obviously give students practice and improve their skills with fundamental mathematical operations. However, such calculations are probably carried out only rarely by hand in the real world. Asking students to calculate both by hand and with a calculator helps provide them with a personal check on their own work.

When calculators are used, encourage students to develop algorithms that they can use repeatedly with problems of a single type. For example, one way of completing Mrs. Van Dusen's payroll is to calculate the number of hours worked each day (by inspection) and add the number of hours worked for the first week (manually or by inspection). That total can then be multiplied by the hourly wage and entered into memory (M+). The procedure can then be repeated for the second week, with the answer added to memory also. The answer in memory (MR) can then be used to calculate federal income tax and F.I.C.A. and Medicare taxes. Total deductions can then be subtracted from memory (M–) to give net pay.

Answers to Text Questions, pages 2 and 6

1. A *wage* is money paid to an employee for work done, usually figured on an hourly, daily, or piece-work basis. Distinguished from that is a *salary,* which is a fixed payment at regular intervals for services, usually other than manual or mechanical labor: A construction worked receives a *wage* of $37.85 an hour. A banker receives a *salary* of $98,500 a year.

2. *Gross pay* is the total pay earned by an employee before deductions. *Net pay* is the amount of wage or salary left after deductions have been taken out of gross pay.

3. Deductions required by law are federal and state income taxes. F.I.C.A. (Social Security) and Medicare contributions are required from all employees except state and federal employees who have their own pension or retirement systems. Examples of other possible deductions include health insurance payments, contributions to charitable organizations or campaigns, profit-sharing plans, union dues, tax-sheltered annuities or other retirement programs, and direct deposits (to employees' savings or checking accounts).

4. Student answers will vary.

5. Some possible reasons include overtime; loss of pay due to illness; change in numbers of exemptions; change in rate of medical costs; change in rate of salary or wage payments; change in other deductions; deductions made only once a month and not every week.

6. An *exemption* (also called an *allowance*) is a certain amount of money or a number of persons that can be subtracted from a person's income tax. For example, a person can claim himself or herself, his or her spouse, and their children as exemptions. A woman who is head of a household with three dependent children can claim four exemptions (herself plus three children) on her income tax. The more exemptions a person has, the less income tax is withheld from his or her paycheck.

7. Social Security tax = $\$452 \times 12.4\% = \$56.05$
    Medicare tax = $\$452 \times 2.9\% = \$13.11$.

Comments on Internet Activities, pages 2, 4, and 6

[**Note:** The information on many Internet sites changes. In many cases, sites become obsolete or are discontinued. The answers provided below are taken from sites believed to be reliable and likely to remain in existence.]

1. The *minimum wage law* is the common name given to the Fair Labor Standards Act. The act provides that persons employed in certain occupations may receive *no less than* certain standard wages. Congress has written and amended the act to cover as wide a range of employees as possible. Essentially, any person engaged in any form of interstate commerce is covered. For example, any office worker who handles mail or makes telephone calls is covered by the minimum wage law. The law itself is quite complex with many specific provisions. Relevant web sites include the following:

    > FLSA Advisor
    > http://www.dol.gov/elaws/flsa.htm
    > Minimum Wage Laws in the states (state laws)
    > http://www.dol.gov/dol/esa/public/minwage/america.htm
    > The Minimum Wage (federal law)
    > http://www.dol.gov/dol/topic/wages/minimumwage.htm

2. Federal law requires that men and women who do the same job must receive the same pay. In spite of this law, pay scales for men and women are often quite different, even when they are performing almost exactly the same work. For more information, see the following web sites:

    > Equal Employment Opportunities Commission (EEOC)
    > http://www.eeoc.gov
    > NOLO Law for All
    > Topic: Equal Pay for Equal Work
    > http://www.nolo.com/
    > Lycos
    > Topic: Your Federal Liability
    > http://www.lycos.com/business

3. A Social Security number is a nine-digit number assigned to most Americans by the Social Security Administration. The primary use of the Social Security number is for the identification of persons for the payment of federal income tax, F.I.C.A., Medicare, and other federal programs. Over the past few decades, it has become much more widely used as an identifier for many nongovernmental programs, such as the use of credit cards and banking programs. A Social Security number is required for every person over the age of one year whose parents declare that child as a dependent on their federal income tax. For further information on Social Security numbers, see

    > Social Security Online
    > http://www.ssa.gov
    > http://www.ssa.gov/history/

4. An *IRA* account is an Individual Retirement Account. It is one of the mechanisms permitted by U.S. tax laws for people to set aside money for their retirement. Other types of accounts that students may find include 401(k) plans; Keough plans; simplified employee pension plans; and Roth IRAs. These plans will be discussed in greater detail later in the book.

5. Federal income tax tables can be found on the Internal Revenue Service web site at http://www.irs.gov/prod/ind_info/tax_tables/index.html

6. State income tax tables can be found on state government web sites. A general source is the Federation of Tax Administrators, located at http://www.taxadmin.org/fta/forms.ssi

Student Page 7

BULL & BEARE STOCK ADVISORS
17 Investment Blvd.
Grand View, MN 55550

ROSCOE VAN DUSEN	368-34-5686	048-377	02	January 3
NAME	SOC. SEC. NO.	EMPLOYEE NO.	EXEMPTIONS	PAY PERIOD ENDING

EARNINGS	AMOUNT	DEDUCTIONS	AMOUNT
Weekly Salary	$1,275.96	Federal Income Tax	$180.00
		State Income Tax	$ 34.50*
		F.I.C.A.	$ 79.11
		Medicare	$ 18.50
		Health Insurance	$134.88
		Professional Dues	$ 35.00
		IRA	$ 75.00
		United Fund	$ 25.00

GROSS EARNINGS	TOTAL DEDUCTIONS	NET PAY
$1,275.96	$581.99	$693.97

* This figure will vary by state.

Student Page 8

BULL & BEARE STOCK ADVISORS
17 Investment Blvd.
Grand View, MN 55550

ROSCOE VAN DUSEN	368-34-5686	048-377	02	January 10
NAME	SOC. SEC. NO.	EMPLOYEE NO.	EXEMPTIONS	PAY PERIOD ENDING

EARNINGS	AMOUNT	DEDUCTIONS	AMOUNT
Weekly Salary	$1,275.96	Federal Income Tax	$180.00
		State Income Tax	$ 34.50*
		F.I.C.A.	$ 79.11
		Medicare	$ 18.50
		Health Insurance	$134.88
		Professional Dues	$ 35.00
		IRA	$ 75.00
		United Fund	$ 25.00

GROSS EARNINGS	TOTAL DEDUCTIONS	NET PAY
$1,275.96	$581.99	$693.97

* This figure will vary by state.

Student Page 9

BULL & BEARE STOCK ADVISORS
17 Investment Blvd.
Grand View, MN 55550

ROSCOE VAN DUSEN	368-34-5686	048-377	02	January 17
NAME	SOC. SEC. NO.	EMPLOYEE NO.	EXEMPTIONS	PAY PERIOD ENDING

EARNINGS	AMOUNT	DEDUCTIONS	AMOUNT
Weekly Salary	$1,275.96	Federal Income Tax	$180.00
		State Income Tax	$ 34.50*
		F.I.C.A.	$ 79.11
		Medicare	$ 18.50
		Health Insurance	$134.88
		Professional Dues	$ 35.00
		IRA	$ 75.00
		United Fund	$ 25.00

GROSS EARNINGS	TOTAL DEDUCTIONS	NET PAY
$1,275.96	$581.99	$693.97

* This figure will vary by state.

Student Page 10

BULL & BEARE STOCK ADVISORS
17 Investment Blvd.
Grand View, MN 55550

ROSCOE VAN DUSEN	368-34-5686	048-377	02	January 24
NAME	SOC. SEC. NO.	EMPLOYEE NO.	EXEMPTIONS	PAY PERIOD ENDING

EARNINGS	AMOUNT	DEDUCTIONS	AMOUNT
Weekly Salary	$1,275.96	Federal Income Tax	$180.00
		State Income Tax	$ 34.50*
		F.I.C.A.	$ 79.11
		Medicare	$ 18.50
		Health Insurance	$134.88
		Professional Dues	$ 35.00
		IRA	$ 75.00
		United Fund	$ 25.00

GROSS EARNINGS	TOTAL DEDUCTIONS	NET PAY
$1,275.96	$581.99	$693.97

* This figure will vary by state.

Student Page 11

BULL & BEARE STOCK ADVISORS
17 Investment Blvd.
Grand View, MN 55550

ROSCOE VAN DUSEN	368-34-5686	048-377	02	*January 31*
NAME	SOC. SEC. NO.	EMPLOYEE NO.	EXEMPTIONS	PAY PERIOD ENDING

EARNINGS	AMOUNT	DEDUCTIONS	AMOUNT
Weekly Salary	$1,275.96	Federal Income Tax	$180.00
		State Income Tax	$ 34.50*
		F.I.C.A.	$ 79.11
		Medicare	$ 18.50
		Health Insurance	$134.88
		Professional Dues	$ 35.00
		IRA	$ 75.00
		United Fund	$ 25.00

GROSS EARNINGS	TOTAL DEDUCTIONS	NET PAY
$1,275.96	$581.99	$693.97

* This figure will vary by state.

Student Page 13

GRAND VIEW PUBLIC SCHOOLS
Schoolhouse Road
Grand View, MN 55550

PAYROLL RECORD

Employee Name: VIVIAN VAN DUSEN   Social Security No: 368-33-5851
School Assignment: CONGRESS MIDDLE SCHOOL   Exemptions: 3

For Pay Period Ending:: *January 26*

DATE	TIME IN	TIME OUT	HOURS
1/15	8:00	4:00	8
1/16	8:30	4:00	7.5
1/17	8:30	4:00	7.5
1/18	9:00	3:00	6
1/19	9:00	12:00	3
		Total Hours, This Week:	32
		Total Pay, This Week:	$784.00
1/22	8:00	12:00	4
1/23	8:00	12:00	4
1/24	10:00	12:00	2
1/25	9:00	4:00	7
1/26	8:00	4:30	8.5
		Total Hours, This Week:	25.5
		Total Pay, This Week:	$624.75

TOTAL GROSS PAY	FED. TAX	STATE TAX	F.I.C.A.	MEDICARE	TOTAL DEDUCTIONS	NET PAY
$1,408.75	$124.00	$17.90*	$87.34	$20.43	$249.67	$1,159.08

* This figure will vary by state.

Student Page 12

BULL & BEARE STOCK ADVISORS
17 Investment Blvd.
Grand View, MN 55550

ROSCOE VAN DUSEN	368-34-5686	048-377	02	*February 7*
NAME	SOC. SEC. NO.	EMPLOYEE NO.	EXEMPTIONS	PAY PERIOD ENDING

EARNINGS	AMOUNT	DEDUCTIONS	AMOUNT
Weekly Salary	$1,275.96	Federal Income Tax	$180.00
		State Income Tax	$ 34.50*
		F.I.C.A.	$ 79.11
		Medicare	$ 18.50
		Health Insurance	$134.88
		Professional Dues	$ 35.00
		IRA	$ 75.00
		United Fund	$ 25.00

GROSS EARNINGS	TOTAL DEDUCTIONS	NET PAY
$1,275.96	$581.99	$693.97

* This figure will vary by state.

Student Page 14

GRAND VIEW PUBLIC SCHOOLS
Schoolhouse Road
Grand View, MN 55550

PAYROLL RECORD

Employee Name: VIVIAN VAN DUSEN   Social Security No: 368-33-5851
School Assignment: CONGRESS MIDDLE SCHOOL   Exemptions: 3

For Pay Period Ending:: *February 9*

DATE	TIME IN	TIME OUT	HOURS
1/29	8:00	4:00	8
1/30	7:30	4:30	9
1/31	7:30	4:30	9
2/1	8:00	4:00	8
2/2	8:00	4:00	8
		Total Hours, This Week:	42
		Total Pay, This Week:	$1,053.50
2/5	8:30	12:00	3.5
2/6	8:00	12:30	4.5
2/7	9:00	12:00	3
2/8	9:00	4:00	7
2/9	8:00	4:30	8.5
		Total Hours, This Week:	26.5
		Total Pay, This Week:	$649.25

TOTAL GROSS PAY	FED. TAX	STATE TAX	F.I.C.A.	MEDICARE	TOTAL DEDUCTIONS	NET PAY
$1,702.75	$169.00	$21.30*	$105.57	$24.69	$320.56	$1,382.19

* This figure will vary by state.

Student Page 15

GRAND VIEW PUBLIC SCHOOLS
Schoolhouse Road
Grand View, MN 55550

PAYROLL RECORD

Employee Name: **VIVIAN VAN DUSEN**   Social Security No: **368-33-5851**
School Assignment: **CONGRESS MIDDLE SCHOOL**   Exemptions: **3**

For Pay Period Ending: _February 23_

DATE	TIME IN	TIME OUT	HOURS
2/12	8:00	4:00	8
2/13	8:30	4:30	8
2/14	8:00	4:30	8.5
2/15	8:00	5:30	9.5
2/16	8:00	5:00	9
		Total Hours, This Week:	43
		Total Pay, This Week:	$1,090.25
2/19	8:00	12:00	4
2/20	8:00	12:30	4.5
2/21	10:00	12:00	2
2/22	9:00	3:00	6
2/23	9:00	4:30	7.5
		Total Hours, This Week:	24
		Total Pay, This Week:	$588.00

TOTAL GROSS PAY	FED. TAX	STATE TAX	F.I.C.A.	MEDICARE	TOTAL DEDUCTIONS	NET PAY
$1,678.25	$163.00	$20.90*	$104.05	$24.33	$312.28	$1,365.97

* This figure will vary by state.

Student Page 16

GRAND VIEW PUBLIC SCHOOLS
Schoolhouse Road
Grand View, MN 55550

PAYROLL RECORD

Employee Name: **VIVIAN VAN DUSEN**   Social Security No: **368-33-5851**
School Assignment: **CONGRESS MIDDLE SCHOOL**   Exemptions: **3**

For Pay Period Ending: _March 9_

DATE	TIME IN	TIME OUT	HOURS
2/26	8:30	4:30	8
2/27	8:00	3:00	7
2/28	9:30	4:00	6.5
3/1	9:00	2:00	5
3/2	9:00	12:00	3
		Total Hours, This Week:	29.5
		Total Pay, This Week:	$722.75
3/5	9:00	12:00	3
3/6	8:00	12:00	4
3/7	10:00	3:00	5
3/8	9:00	3:00	6
3/9	8:00	4:30	8.5
		Total Hours, This Week:	26.5
		Total Pay, This Week:	$649.25

TOTAL GROSS PAY	FED. TAX	STATE TAX	F.I.C.A.	MEDICARE	TOTAL DEDUCTIONS	NET PAY
$1,372.00	$118.00	$17.90*	$85.06	$19.89	$242.85	$1,131.15

* This figure will vary by state.

Student Page 17

GRAND VIEW PUBLIC SCHOOLS
Schoolhouse Road
Grand View, MN 55550

PAYROLL RECORD

Employee Name: **VIVIAN VAN DUSEN**   Social Security No: **368-33-5851**
School Assignment: **CONGRESS MIDDLE SCHOOL**   Exemptions: **3**

For Pay Period Ending: _March 23_

DATE	TIME IN	TIME OUT	HOURS
3/12	8:00	4:00	8
3/13	9:30	4:30	7
3/14	9:00	8:00	11
3/15	9:00	6:00	9
3/16	7:00	5:30	10.5
		Total Hours, This Week:	45.5
		Total Pay, This Week:	$1,182.13
3/19	9:00	3:00	6
3/20	8:00	12:00	4
3/21	10:00	1:00	3
3/22	9:00	4:00	7
3/23	8:00	4:30	8.5
		Total Hours, This Week:	28.5
		Total Pay, This Week:	$698.25

TOTAL GROSS PAY	FED. TAX	STATE TAX	F.I.C.A.	MEDICARE	TOTAL DEDUCTIONS	NET PAY
$1,880.38	$196.00	$23.80*	$116.58	$27.27	$363.65	$1,516.73

* This figure will vary by state.

Student Page 18

GRAND VIEW PUBLIC SCHOOLS
Schoolhouse Road
Grand View, MN 55550

PAYROLL RECORD

Employee Name: **VIVIAN VAN DUSEN**   Social Security No: **368-33-5851**
School Assignment: **CONGRESS MIDDLE SCHOOL**   Exemptions: **3**

For Pay Period Ending: _April 6_

DATE	TIME IN	TIME OUT	HOURS
3/26	8:30	4:00	7.5
3/27	8:30	4:00	7.5
3/28	8:30	4:30	8
3/29	9:00	3:30	6.5
3/30	9:00	12:30	3.5
		Total Hours, This Week:	33
		Total Pay, This Week:	$808.50
4/2	8:30	12:30	4
4/3	8:30	12:30	4
4/4	10:00	2:00	4
4/5	9:00	4:00	7
4/6	8:30	4:00	7.5
		Total Hours, This Week:	26.5
		Total Pay, This Week:	$649.25

TOTAL GROSS PAY	FED. TAX	STATE TAX	F.I.C.A.	MEDICARE	TOTAL DEDUCTIONS	NET PAY
$1,457.75	$130.00	$17.90*	$90.38	$21.14	$259.42	$1,198.33

* This figure will vary by state.

Student Page 19

SWIFTIE'S SERVICE
140 Mandrell Ave.
Grand View, MN 55550

PAYROLL RECORD

Employee: **CRAIG VAN DUSEN**   S. S. Number: **246-38-0150**   Exemptions: **0**

Pay Period: From ___1/8___ to ___1/12___

Regular Hours: ___40___   Regular Pay: $ ___250.00___
Overtime Hours: ___8___   Overtime Pay: $ ___75.04___

Extra Jobs:

___3___	flats changed @ $8.00	=	$ ___24.00___
___2___	road service @ $15.00	=	$ ___30.00___
___4___	car washes @ $15.00	=	$ ___60.00___
___2___	car waxes @ $27.50	=	$ ___55.00___
___5___	hrs repair work @ $10.50/hr (apprentice pay)	=	$ ___52.50___
		Total Extras:	$ ___221.50___

TOTAL GROSS PAY	FED. TAX	STATE TAX	F.I.C.A.	MEDICARE	TOTAL DEDUCTIONS	NET PAY
$546.54	$74.00	$5.10*	$33.89	$7.92	$120.91	$425.63

* This figure will vary by state.

Student Page 21

SWIFTIE'S SERVICE
140 Mandrell Ave.
Grand View, MN 55550

PAYROLL RECORD

Employee: **CRAIG VAN DUSEN**   S. S. Number: **246-38-0150**   Exemptions: **0**

Pay Period: From ___1/22___ to ___1/26___

Regular Hours: ___36.5___   Regular Pay: $ ___228.13___
Overtime Hours: ___0___   Overtime Pay: $ ___0___

Extra Jobs:

___0___	flats changed @ $8.00	=	$ ___0___
___0___	road service @ $15.00	=	$ ___0___
___0___	car washes @ $15.00	=	$ ___0___
___0___	car waxes @ $27.50	=	$ ___0___
___2___	hrs repair work @ $10.50/hr (apprentice pay)	=	$ ___21.00___
		Total Extras:	$ ___21.00___

TOTAL GROSS PAY	FED. TAX	STATE TAX	F.I.C.A.	MEDICARE	TOTAL DEDUCTIONS	NET PAY
$249.12	$29.00	$1.10*	$15.45	$3.61	$49.16	$199.97

* This figure will vary by state.

Student Page 20

SWIFTIE'S SERVICE
140 Mandrell Ave.
Grand View, MN 55550

PAYROLL RECORD

Employee: **CRAIG VAN DUSEN**   S. S. Number: **246-38-0150**   Exemptions: **0**

Pay Period: From ___1/15___ to ___1/19___

Regular Hours: ___40___   Regular Pay: $ ___250.00___
Overtime Hours: ___2___   Overtime Pay: $ ___18.76___

Extra Jobs:

___2___	flats changed @ $8.00	=	$ ___16.00___
___0___	road service @ $15.00	=	$ ___0___
___2___	car washes @ $15.00	=	$ ___30.00___
___2___	car waxes @ $27.50	=	$ ___55.00___
___4___	hrs repair work @ $10.50/hr (apprentice pay)	=	$ ___42.00___
		Total Extras:	$ ___143.00___

TOTAL GROSS PAY	FED. TAX	STATE TAX	F.I.C.A.	MEDICARE	TOTAL DEDUCTIONS	NET PAY
$411.76	$55.00	$1.60*	$25.53	$5.97	$91.10	$320.66

* This figure will vary by state.

Student Page 22

SWIFTIE'S SERVICE
140 Mandrell Ave.
Grand View, MN 55550

PAYROLL RECORD

Employee: **CRAIG VAN DUSEN**   S. S. Number: **246-38-0150**   Exemptions: **0**

Pay Period: From ___1/29___ to ___2/2___

Regular Hours: ___17.5___   Regular Pay: $ ___109.38___
Overtime Hours: ___0___   Overtime Pay: $ ___0___

Extra Jobs:

___0___	flats changed @ $8.00	=	$ ___0___
___0___	road service @ $15.00	=	$ ___0___
___3___	car washes @ $15.00	=	$ ___45.00___
___1___	car waxes @ $27.50	=	$ ___27.50___
___2___	hrs repair work @ $10.50/hr (apprentice pay)	=	$ ___21.00___
		Total Extras:	$ ___93.50___

TOTAL GROSS PAY	FED. TAX	STATE TAX	F.I.C.A.	MEDICARE	TOTAL DEDUCTIONS	NET PAY
$202.88	$23.00	$1.10*	$12.58	$2.94	$39.62	$163.26

* This figure will vary by state.

8   *Math in Everyday Life*

Student Page 23

Swiftie's Service
140 Mandrell Ave.
Grand View, MN 55550

PAYROLL RECORD

Employee: **CRAIG VAN DUSEN**   S. S. Number: **246-38-0150**   Exemptions: **0**

Pay Period: From _____ 2/5 _____ to _____ 2/9 _____

Regular Hours: _____ 40 _____   Regular Pay: $ _____ 250.00 _____
Overtime Hours: _____ 3.5 _____   Overtime Pay: $ _____ 32.81 _____

Extra Jobs:

3	flats changed @ $8.00	=	$	24.00
2	road service @ $15.00	=	$	30.00
2	car washes @ $15.00	=	$	30.00
2	car waxes @ $27.50	=	$	55.00
3	hrs repair work @ $10.50/hr (apprentice pay)	=	$	31.50
		Total Extras:	$	170.50

TOTAL GROSS PAY	FED. TAX	STATE TAX	F.I.C.A.	MEDICARE	TOTAL DEDUCTIONS	NET PAY
$453.31	$61.00	$5.30*	$28.11	$6.57	$100.98	$352.33

* This figure will vary by state.

Student Page 24

Swiftie's Service
140 Mandrell Ave.
Grand View, MN 55550

PAYROLL RECORD

Employee: **CRAIG VAN DUSEN**   S. S. Number: **246-38-0150**   Exemptions: **0**

Pay Period: From _____ 2/12 _____ to _____ 2/16 _____

Regular Hours: _____ 40 _____   Regular Pay: $ _____ 250.00 _____
Overtime Hours: _____ 4.25 _____   Overtime Pay: $ _____ 39.84 _____

Extra Jobs:

1	flats changed @ $8.00	=	$	8.00
1	road service @ $15.00	=	$	15.00
2	car washes @ $15.00	=	$	30.00
1	car waxes @ $27.50	=	$	27.50
1	hrs repair work @ $10.50/hr (apprentice pay)	=	$	10.50
		Total Extras:	$	91.00

TOTAL GROSS PAY	FED. TAX	STATE TAX	F.I.C.A.	MEDICARE	TOTAL DEDUCTIONS	NET PAY
$380.84	$50.00	$4.60*	$23.61	$5.52	$83.73	$297.11

* This figure will vary by state.

Student Page 25

Vanessa's Earnings

Following are the records kept by Vanessa of her baby-sitting earnings. She receives $3.75 per hour for sitting up to midnight and $4.25 per hour after midnight.

Baby-sitting
Vanessa Van Dusen

Date	Job	Hours	Pay
1/3	Ms. Jackson	7:00–11:30	$16.88
1/4	Mr. Washington	8:30–1:30	$19.50
1/10	Mrs. Andronitz	8:00–12:00	$15.00
1/11	Mr. Simpson	7:30–12:30	$19.00
1/12	Ms. Jackson (afternoon)	3:00–4:30	$ 5.63
1/17	Mr. Lee	7:00–1:00	$23.00
1/18	Ms. Hoffer	8:30–11:30	$11.25
1/20	Ms. Jackson	8:00–12:30	$17.13
1/24	Mrs. Lee	7:30–12:00	$16.88
1/25	Mr. Austin	5:00–1:00	$30.52
1/31	Mr. Austin	9:00–1:30	$17.63
2/1	Mr. Simpson	8:00–12:30	$17.13
2/6	Mrs. Andronitz	8:30–1:00	$17.38
2/7	Ms. Hoffer	8:00–1:00	$19.25
2/8	Mr. Austin (afternoon)	2:00–5:00	$11.25
2/21	Ms. Jackson	7:30–12:30	$19.00
2/26	Mr. Simpson	8:00–1:00	$19.25
3/3	Ms. Jackson	9:30–2:00	$17.88
3/6	Mrs. Simpson	7:30–1:00	$21.13

Chapter 2

BUDGETS

Teaching Notes

In connection with Exercise IA, Long-Term Budgets: Planning for College, you may want to conduct a discussion of whether saving for a child's college education should come from the parents alone, the child alone, or from some combination of both. What family or economic conditions might help determine this decision? You might also raise a question about planning for college if some doubt exists about wanting to attend. Should money be set aside for college expenses in such a case?

The column labeled Due Date in Exercise IC, Major Annual Expenses, is included so that students can consider these expenses when they make out Monthly Budgets in Exercise II.

Multiple copies of Monthly Budgets are provided so that students can think about the way expenses change at different times of the year. The January budget may include heating costs, for example; the July budget may include an item for the family vacation.

You may want to have a discussion about the Expenses part of the Monthly Budgets in Exercise II. The figures provided here reflect the results of one national survey of spending habits across the nation. These figures may differ from one part of the country to another and from one household to another.

Proportions may vary from one month to another. For example, a family may pay all its taxes in two or three months. The family's budget for taxes in those months might be 15–25% of the total budget *for that month*. In other months, however, the proportion set aside for taxes would then be 0%.

On the other hand, a family might choose to set aside a definite amount each month toward all its expenses. As an example, many families now set aside a certain amount each month, summer and winter, toward their heating bills. In this case, the column headed Average Part of the Family Budget would be a good guideline for students who need to know about how much the family will spend overall on each item in the year.

Answers to Text Questions, pages 27–28

1. Some possible advantages of a budget include the following:
   (a) Helps a family keep spending in line with income
   (b) Helps a family compare expenses so that essentials are taken care of before luxuries are paid for
   (c) Helps a family plan for future spending to ensure that money will be available when those future expenses are incurred

   Most people do not think that there are any disadvantages to having a budget.

2. *Short-term* budgets are those used to plan for expenses that will occur within a relatively brief time period, usually a month. *Long-term* budgets are used for year-long planning and for major expenses that must be planned over a longer period of time.

3. In order to make out a monthly budget, one needs to know all income and all expenses during the period and what percentages of income should be allocated to each expenditure.

4. Additional income that could be used in paying for a college education might come from scholarships, grants, or loans. Outright gifts of money (for example, scholarships) need not be included in the budget except to decrease the amount of money that must be obtained from other sources. Loans for college expenses often do not have to be paid back until after graduation. Payments coming due at that time should be included in a long-term budget.

5. Some of the bills one might expect to have in December, but not in June, would be heating bills; cost of heavy winter clothing; higher

electrical bills (depending upon whether the family uses air-conditioning in the summer); having the driveway plowed; and certain school expenses (if school ends in June). June expenses might include costs of gardening and yard work and the cost of the family's summer vacation.

6. In planning January's budget, the family makes some guesses about how they will spend their money and what proportion of their income each expenditure will receive. At the end of January, the family can see how closely the actual expenses matched their predicted expenses. This comparison then gives them a more reliable basis for making predictions about their February budget.

7. Answers will vary. Some students may point out that parents have some responsibility for paying for the upbringing of their children. Others may mention that students who pay for their own education may take that education more seriously. Some students may find it difficult to work and go to school at the same time. Parental help with expenses can be very important. However, parents may simply not have the money to set aside for a child's future college education. Thus, the child may have to take much of the financial responsibility for his or her education.

Comments on Internet Activities, page 28

1. Information about these topics can be found at the following web sites:

> The Art of Savvy Budgeting (From *Black Enterprise* magazine)
> http://www.findarticles.com/ m1365/10_30/61834965/p1/ article.jhtml

Consumer Alert
Topic: Commonsense Consumer Column
http://www.consumeralert.org/
FREE Personal Budgeting Help for Home and Business
http://www.budgetresource.com/
A Sample Household Budget
www.dacomp.com/sample.html

2. Students are likely to find two kinds of information about college tuition on the Internet. First, tuition rates (along with other expenses) are often listed on the web sites of specific colleges and universities. Second, articles frequently appear in the popular press updating tuition charges for many types of institutions.

3. The data provided in the student workbook was obtained from the U.S. Department of Energy's Office of Building Technology, State and Community Programs' *Core Databook* at http://208.226.167.195/btscore98/ xlsfiles/5.10.18.htm. Other web sites may contain similar information on specific classes of appliances.

4. Consumers are making more and more purchases via the Internet today. This approach to shopping is not equally useful for all types of products, however. Students will find a number of web sites through which they can buy some appliances listed in this workbook; others are harder to find. You might discuss with students the pros and cons of shopping on-line, with particular attention to the types of products for which this method of shopping is more reasonable. You might also discuss with them alternative approaches to shopping for major appliances when the web is not a satisfactory avenue.

Completed Tables (pp. 29–31)

Exercise I. Long-Term Budgets

A. Planning for College

Child	Age Now	Age When Entering College	Years to College	Projected Cost of College	Current Savings	Additional Savings Needed	Additional Savings Needed per Year
Craig	17	18	1	$30,000	$22,000	$ 8,000	$8,000.00
Vanessa	13	18	5	$36,000	$2,400	$33,600	$6,720.00
Charley	6	18	12	$48,000	$1,000	$47,000	$3,916.67

Child	Age Now	Age When Entering College	Years to College	Projected Cost of College	Current Savings	Additional Savings Needed	Additional Savings Needed per Year
John	16	18	2	$32,000	$28,000	$ 4,000	$2,000.00
Elizabeth	12	18	6	$40,000	$12,000	$28,000	$4,666.67
Rose	9	18	9	$55,000	$4,000	$51,000	$5,666.67

B. Future Appliance Expenses

Item	Years Since Purchase	Average Number of Years Appliance Will Last	Years Until Appliance Must Be Replaced	Average Cost of New Appliance	Amount to Be Set Aside Each Year
Refrigerator	12	16	4	$585	$146.25
Electric Range	9	16	7	$495	$70.71
Washing Machine	10	11	1	$315	$315.00
Dryer	10	14	4	$325	$81.25
Vacuum Cleaner	5	15	10	$135	$13.50
Sewing Machine	8	24	16	$275	$17.19
Television	2	11	9	$515	$57.22
Automobile	3	6	3	$18,500	$6,166.67
Personal Computer	2	7	5	$2,475	$495.00

Item	Years Since Purchase	Average Number of Years Appliance Will Last	Years Until Appliance Must Be Replaced	Average Cost of New Appliance	Amount to Be Set Aside Each Year
Refrigerator	3	16	13	$585	$45.00
Electric Range	5	16	11	$495	$45.00
Washing Machine	8	11	3	$315	$105.00
Dryer	3	14	11	$325	$29.55
Vacuum Cleaner	9	15	6	$135	$22.50
Sewing Machine	6	24	18	$275	$15.28
Television	4	11	7	$515	$73.57
Automobile	2	6	4	$18,500	$4,625.00
Personal Computer	3	7	4	$3,125	$781.25

C. Major Annual Expenses

Item	Annual Cost	Average Monthly Cost[1]	Average Weekly Cost[2]	Due Date
Federal Income Tax[3]	$1,781	$148.42	$34.25	April 15
State Income Tax[3]	$652	$54.33	$12.54	April 15
Life Insurance	$895	$74.58	$17.21	November 5
Property Taxes	$2,398	$199.83	$46.12	June 30
House Insurance	$677	$56.42	$13.02	August 1

[1]Based on 12 months   [2]Based on 52 weeks   [3]Above and beyond deductions; based on last year's payments

Student Page 33

Example

Budget for the month of _____January_____

NET INCOME

Mr. Van Dusen	$2,770
Mrs. Van Dusen	$2,580
Craig*	$1,000
Vanessa*	$ 100
Other income	$ 100
TOTAL	$ 6,550

*Decide how much Craig and Vanessa should contribute to family expenses this month.

EXPENSES

ITEM	AVERAGE PART OF FAMILY BUDGET[1]	AMOUNT SET ASIDE FOR MONTH	AMOUNT SET ASIDE FOR WEEK[2]
Food	15%	$979.50	$244.88
Mortgage	19%	$1,240.70	$310.18
Utilities	6%	$391.80	$97.95
Clothing	5%	$326.50	$81.63
Personal Care	1%	$65.30	$16.33
Health Care	5%	$326.50	$81.63
Automobile	12%	$783.60	$195.90
Entertainment	4%	$261.20	$65.30
Gifts, Contributions	3%	$195.90	$48.98
Insurance (all kinds)	7%	$457.10	$114.28
Taxes (all kinds)	9%	$587.70	$146.93
Retirement	6%	$391.80	$97.95
Savings (all kinds)	5%	$326.50	$81.63
Other Expenses	3%	$195.90	$48.98

[1]Based on national survey of family expenses in the United States   [2]Figured on the basis of four weeks per month

Student Page 35

Answers will vary.

Budget for the month of _____July_____

NET INCOME

Mr. Van Dusen	_____
Mrs. Van Dusen	_____
Craig*	_____
Vanessa*	_____
Other income	_____
TOTAL	$ _____

*Decide how much Craig and Vanessa should contribute to family expenses this month.

EXPENSES

ITEM	AVERAGE PART OF FAMILY BUDGET[1]	AMOUNT SET ASIDE FOR MONTH	AMOUNT SET ASIDE FOR WEEK[2]
Food	15%		
Mortgage	19%		
Utilities	6%		
Clothing	5%		
Personal Care	1%		
Health Care	5%		
Automobile	12%		
Entertainment	4%		
Gifts, Contributions	3%		
Insurance (all kinds)	7%		
Taxes (all kinds)	9%		
Retirement	6%		
Savings (all kinds)	5%		
Other Expenses	3%		

[1]Based on national survey of family expenses in the United States   [2]Figured on the basis of four weeks per month

Student Page 34

Answers will vary.

Budget for the month of _____April_____

NET INCOME

Mr. Van Dusen	_____
Mrs. Van Dusen	_____
Craig*	_____
Vanessa*	_____
Other income	_____
TOTAL	$ _____

*Decide how much Craig and Vanessa should contribute to family expenses this month.

EXPENSES

ITEM	AVERAGE PART OF FAMILY BUDGET[1]	AMOUNT SET ASIDE FOR MONTH	AMOUNT SET ASIDE FOR WEEK[2]
Food	15%		
Mortgage	19%		
Utilities	6%		
Clothing	5%		
Personal Care	1%		
Health Care	5%		
Automobile	12%		
Entertainment	4%		
Gifts, Contributions	3%		
Insurance (all kinds)	7%		
Taxes (all kinds)	9%		
Retirement	6%		
Savings (all kinds)	5%		
Other Expenses	3%		

[1]Based on national survey of family expenses in the United States   [2]Figured on the basis of four weeks per month

Student Page 36

Answers will vary.

Budget for the month of _____November_____

NET INCOME

Mr. Van Dusen	_____
Mrs. Van Dusen	_____
Craig*	_____
Vanessa*	_____
Other income	_____
TOTAL	$ _____

*Decide how much Craig and Vanessa should contribute to family expenses this month.

EXPENSES

ITEM	AVERAGE PART OF FAMILY BUDGET[1]	AMOUNT SET ASIDE FOR MONTH	AMOUNT SET ASIDE FOR WEEK[2]
Food	15%		
Mortgage	19%		
Utilities	6%		
Clothing	5%		
Personal Care	1%		
Health Care	5%		
Automobile	12%		
Entertainment	4%		
Gifts, Contributions	3%		
Insurance (all kinds)	7%		
Taxes (all kinds)	9%		
Retirement	6%		
Savings (all kinds)	5%		
Other Expenses	3%		

[1]Based on national survey of family expenses in the United States   [2]Figured on the basis of four weeks per month

Answers will vary.

Student Page 38

ITEM	JAN	FEB	MAR	APR	MAY	JUNE
Food[1]	$	$	$	$	$	$
Mortgage[1]	$	$	$	$	$	$
Household Expenses[1]	$	$	$	$	$	$
Major Appliances[2]	$	$	$	$	$	$
Property Taxes[3]	$	$	$	$	$	$
Insurance[3]	$	$	$	$	$	$
Clothing[1]	$	$	$	$	$	$
Personal Care[1]	$	$	$	$	$	$
Health Care[1]	$	$	$	$	$	$
Entertainment[1]	$	$	$	$	$	$
Automobile[1]	$	$	$	$	$	$
Income Taxes[3]	$	$	$	$	$	$
Education Savings[4]	$	$	$	$	$	$
Savings (other)[1]	$	$	$	$	$	$
Gifts and Contributions[1]	$	$	$	$	$	$

[1]Budget the same amount every month, if possible. See Exercise II. Short-Term Budgets.
[2]See Exercise IB: Future Appliance Expenses.
[3]See Exercise IC: Major Annual Expenses.
[4]See Exercise IA: Planning for College.

Answers will vary.

Student Page 39

ITEM	JULY	AUG	SEP	OCT	NOV	DEC
Food[1]	$	$	$	$	$	$
Mortgage[1]	$	$	$	$	$	$
Household Expenses[1]	$	$	$	$	$	$
Major Appliances[2]	$	$	$	$	$	$
Property Taxes[3]	$	$	$	$	$	$
Insurance[3]	$	$	$	$	$	$
Clothing[1]	$	$	$	$	$	$
Personal Care[1]	$	$	$	$	$	$
Health Care[1]	$	$	$	$	$	$
Entertainment[1]	$	$	$	$	$	$
Automobile[1]	$	$	$	$	$	$
Income Taxes[3]	$	$	$	$	$	$
Education Savings[4]	$	$	$	$	$	$
Savings (other)[1]	$	$	$	$	$	$
Gifts and Contributions[1]	$	$	$	$	$	$

[1]Budget the same amount every month, if possible. See Exercise II. Short-Term Budgets.
[2]See Exercise IB: Future Appliance Expenses.
[3]See Exercise IC: Major Annual Expenses.
[4]See Exercise IA: Planning for College.

Chapter 3

CHECKING ACCOUNTS

Answers to Text Questions, pages 41–42

1. To *reconcile* a checking account means to bring into balance the total in a personal checkbook with the balance reported by the bank in the current statement. The reverse side of the bank statement generally explains in detail how to reconcile the statement and may provide work space to perform the necessary calculations.

2. These numbers are strange looking because they are a special type of code printed in magnetic ink. The numbers make it possible for a bank to sort checks by machine, rather than by hand. The numbers identify the issuing bank, the customer account number, and the check number.

3. The name following the expression *Pay to the order of* on a check indicates the person or business to whom payment is made and who may cash the check.

4. In a number such as $\frac{21\text{-}458}{146}$, the upper left number identifies the city or state in which the issuing bank is located. The upper right number identifies the specific bank issuing the check. The bottom number is a Federal Reserve Bank number used in sorting checks. The number as a whole is assigned to a bank by the American Bankers Association to speed up the processing and sorting of checks.

5. When two names appear on a check, it means that either person is permitted to sign the check.

6. Among the precautions to take in writing a check are the following:

    (a) Use ink to fill out the check.

    (b) Destroy all checks on which an error has been made. If the faulty check might be redeemed at the bank for credit, write the word **VOID** across the face of the check.

    (c) Use the same signature that you wrote on the bank's signature card when you opened the account.

    (d) Do not leave any blank spaces on the check.

    (e) Do not leave any empty spaces between the $ sign and the numbers written on the check. Do not leave a blank space before the first word when you write the amount of the check on the next line. These precautions will prevent anyone else from altering the amount of the check.

7. (a) A *debit memo* is a sum charged by a bank to an account for a particular reason, such as a check returned for insufficient funds and/or the bank fee for such a check.

    (b) A *service charge* is a charge made by a bank for a routine cost involved in operating a checking account, such as the cost of making a deposit, writing a check, or maintaining the account for a month.

    (c) A *correcting entry* is the amount that a bank charges or credits an account because of a mathematical error the bank may have made on the preceding statement.

8. The acronym ATM stands for *automated teller machine*. An ATM can accept deposits, pay out cash, report account balances, and perform other simple banking functions without the need for a human teller.

With ATMs, simple banking procedures can be conducted more quickly and more easily than with a human teller. In some cases, they can be used in drive-through banks without the customers ever leaving the car.

ATMs can be beneficial to banks, because they reduce the necessity of hiring human tellers and, thus, produce a savings on employment costs.

Using an ATM may involve some small risk of assault or robbery. Some banks charge customers for using an ATM. Students may suggest other advantages and disadvantages of using ATMs based on their own experiences or those of their families and friends.

9. A signature card is a simple form that a person signs when opening a checking or savings account. The signature shows how a person intends to sign his or her checks. The signature card can be used to verify the authenticity of a check submitted for payment.

10. The term *insufficient funds* means that a person does not have enough money in his or her checking account to cover the cost of one or more checks that have been written against that account. Checks returned for insufficient funds are also known as *bounced checks*.

Comments on Internet Activities, pages 42–43

1. Among the ways in which checking accounts differ are the following:

   (a) There may be a single monthly charge for the account or the account may be free.

   (b) Charges may or may not be assessed for each check written or each deposit made.

   (c) A minimum balance may be required to prevent any or all of the above charges.

   (d) Interest may be paid, provided that some minimum balance remains in the checking account for some period during the month.

   (e) All or some checking charges may be waived if a person is over a specific age or maintains a savings account, loan, or some other type of account with the bank.

Many banks and other financial institutions list the checking accounts they offer on their web sites. They often use clever names to distinguish these accounts from each other. Students should be able to access large, national banks; local and regional banks; and *virtual* banks on the Internet.

2. Searches for terms used in this question will produce a number of web sites that describe on-line banking, virtual banking, and methods for electronic funds transfers. Some examples of such web sites include the following:

   VirtualBank
   http://www.virtualbank.com/

   PayTrust
   http://www.paytrust.com/

   General EFT Information
   http://www.fms.treas.gov/eft/

   PayPal
   http://www.paypal.com/

   Electronic Funds Transfer Association
   http://www.efta.org/

   ePaymentSystems, Inc.
   http://www.epaymentsystems.com/

3. *Overdraft protection* means that a bank will pay a check even if there is insufficient money in the account to cover the check. A bank usually requires that a customer open a savings account linked to the checking account and that there be a minimum balance in that account. If there is not enough money in the account to cover a check, the bank transfers enough money from the linked savings account to cover the amount of the check.

4. *Debit cards* allow a person to withdraw money directly from a checking account to pay bills, get cash, or withdraw money from an ATM.

With a debit card, money is deducted instantaneously from a checking account; with a credit card, there is no payment for service until a bill is sent. A useful Internet source for information on debit cards is provided by the National Consumers League at `http://www.natlconsumersleague.org/debitbro.htm.`

5. Information about money orders can be obtained from a local post office or from the United States Postal Service web site at `http://www.usps.com/`. Search for the term *money order* to find a number of publications on this subject. The USPS is developing electronic and on-line payment services, such as the USPS eBillPay Service, described on its web site.

Student Page 44

Roscoe or Vivian Van Dusen
2331 Grapevine Lane
Grand View, MN 55551
(218) 555-1226
MDL# 358-DL-1298
368
26-5595
321
Jan. 2   20____
PAY TO THE ORDER OF CompuNews Monthly $ 35.00
Thirty-five and - - - - - - - - - - - - - - NO/100 DOLLARS
PALINDROME NATIONAL BANK
One Bankers Lane
Grand Island, MN 55553   (Signature)
                                                SIGNATURE
⑈001516 ⑆321080880⑆ 004189048409

Roscoe or Vivian Van Dusen
2331 Grapevine Lane
Grand View, MN 55551
(218) 555-1226
MDL# 358-DL-1298
369
26-5595
321
Jan. 5   20____
PAY TO THE ORDER OF Palindrome National Bank $ 260.00
Two hundred sixty and - - - - - - - - - - NO/100 DOLLARS
PALINDROME NATIONAL BANK
One Bankers Lane
Grand Island, MN 55553   (Signature)
                                                SIGNATURE
⑈001516 ⑆321080880⑆ 004189048409

Roscoe or Vivian Van Dusen
2331 Grapevine Lane
Grand View, MN 55551
(218) 555-1226
MDL# 358-DL-1298
370
26-5595
321
Jan. 6   20____
PAY TO THE ORDER OF Franklin Furniture Depot $ 96.70
Ninety-six and - - - - - - - - - - - - - 70/100 DOLLARS
PALINDROME NATIONAL BANK
One Bankers Lane
Grand Island, MN 55553   (Signature)
                                                SIGNATURE
⑈001516 ⑆321080880⑆ 004189048409

Roscoe or Vivian Van Dusen
2331 Grapevine Lane
Grand View, MN 55551
(218) 555-1226
MDL# 358-DL-1298
371
26-5595
321
Jan. 6   20____
PAY TO THE ORDER OF Golden Honest Insurance Company $ 135.45
One hundred thirty-five and - - - - - - - 45/100 DOLLARS
PALINDROME NATIONAL BANK
One Bankers Lane
Grand Island, MN 55553   (Signature)
                                                SIGNATURE
⑈001516 ⑆321080880⑆ 004189048409

Student Page 45

Roscoe or Vivian Van Dusen
2331 Grapevine Lane
Grand View, MN 55551
(218) 555-1226
MDL# 358-DL-1298
372
26-5595
321
Jan. 6   20____
PAY TO THE ORDER OF Quad States Gas & Electric $ 43.40
Forty-three and - - - - - - - - - - - - 40/100 DOLLARS
PALINDROME NATIONAL BANK
One Bankers Lane
Grand Island, MN 55553   (Signature)
                                                SIGNATURE
⑈001516 ⑆321080880⑆ 004189048409

Roscoe or Vivian Van Dusen
2331 Grapevine Lane
Grand View, MN 55551
(218) 555-1226
MDL# 358-DL-1298
373
26-5595
321
Jan. 6   20____
PAY TO THE ORDER OF A.J. Franklin Testimonial $ 25.00
Twenty-five and - - - - - - - - - - - - NO/100 DOLLARS
PALINDROME NATIONAL BANK
One Bankers Lane
Grand Island, MN 55553   (Signature)
                                                SIGNATURE
⑈001516 ⑆321080880⑆ 004189048409

Roscoe or Vivian Van Dusen
2331 Grapevine Lane
Grand View, MN 55551
(218) 555-1226
MDL# 358-DL-1298
374
26-5595
321
Jan. 11   20____
PAY TO THE ORDER OF Committee to Improve Our Schools $ 50.00
Fifty and - - - - - - - - - - - - - - NO/100 DOLLARS
PALINDROME NATIONAL BANK
One Bankers Lane
Grand Island, MN 55553   (Signature)
                                                SIGNATURE
⑈001516 ⑆321080880⑆ 004189048409

Roscoe or Vivian Van Dusen
2331 Grapevine Lane
Grand View, MN 55551
(218) 555-1226
MDL# 358-DL-1298
375
26-5595
321
Jan. 12   20____
PAY TO THE ORDER OF Pappa' Grocery $ 123.30
One hundred twenty-three and - - - - - 30/100 DOLLARS
PALINDROME NATIONAL BANK
One Bankers Lane
Grand Island, MN 55553   (Signature)
                                                SIGNATURE
⑈001516 ⑆321080880⑆ 004189048409

Student Page 46

Roscoe or Vivian Van Dusen
2331 Grapevine Lane
Grand View, MN 55551
(218) 555-1226
MDL# 358-DL-1298
376
26-5595
321
Jan. 12   20____
PAY TO THE ORDER OF Best & Fitt Clothiers $ 85.00
Eighty-five and - - - - - - - - - - - NO/100 DOLLARS
PALINDROME NATIONAL BANK
One Bankers Lane
Grand Island, MN 55553   (Signature)
                                                SIGNATURE
⑈001516 ⑆321080880⑆ 004189048409

Roscoe or Vivian Van Dusen
2331 Grapevine Lane
Grand View, MN 55551
(218) 555-1226
MDL# 358-DL-1298
377
26-5595
321
Jan. 15   20____
PAY TO THE ORDER OF I. Makum Welle, M.D. $ 62.50
Sixty-two and - - - - - - - - - - - - 50/100 DOLLARS
PALINDROME NATIONAL BANK
One Bankers Lane
Grand Island, MN 55553   (Signature)
                                                SIGNATURE
⑈001516 ⑆321080880⑆ 004189048409

Roscoe or Vivian Van Dusen
2331 Grapevine Lane
Grand View, MN 55551
(218) 555-1226
MDL# 358-DL-1298
378
26-5595
321
Jan. 18   20____
PAY TO THE ORDER OF Cross-Country Airlines $ 232.84
Two hundred thirty-two and - - - - - - 84/100 DOLLARS
PALINDROME NATIONAL BANK
One Bankers Lane
Grand Island, MN 55553   (Signature)
                                                SIGNATURE
⑈001516 ⑆321080880⑆ 004189048409

Roscoe or Vivian Van Dusen
2331 Grapevine Lane
Grand View, MN 55551
(218) 555-1226
MDL# 358-DL-1298
379
26-5595
321
Jan. 18   20____
PAY TO THE ORDER OF Popp-Up Seed Company $ 18.50
Eighteen and - - - - - - - - - - - - 50/100 DOLLARS
PALINDROME NATIONAL BANK
One Bankers Lane
Grand Island, MN 55553   (Signature)
                                                SIGNATURE
⑈001516 ⑆321080880⑆ 004189048409

Student Page 47

Roscoe or Vivian Van Dusen
2331 Grapevine Lane
Grand View, MN 55551
(218) 555-1226
MDL# 358-DL-1298
380
26-5595
321
Jan. 21   20____
PAY TO THE ORDER OF Emma C. Payne, D.V.M. $ 35.40
Thirty-five and - - - - - - - - - - - 40/100 DOLLARS
PALINDROME NATIONAL BANK
One Bankers Lane
Grand Island, MN 55553   (Signature)
                                                SIGNATURE
⑈001516 ⑆321080880⑆ 004189048409

Roscoe or Vivian Van Dusen
2331 Grapevine Lane
Grand View, MN 55551
(218) 555-1226
MDL# 358-DL-1298
381
26-5595
321
Jan. 21   20____
PAY TO THE ORDER OF NACUCOPAT $ 50.00
Fifty and - - - - - - - - - - - - - - NO/100 DOLLARS
PALINDROME NATIONAL BANK
One Bankers Lane
Grand Island, MN 55553   (Signature)
                                                SIGNATURE
⑈001516 ⑆321080880⑆ 004189048409

Roscoe or Vivian Van Dusen
2331 Grapevine Lane
Grand View, MN 55551
(218) 555-1226
MDL# 358-DL-1298
382
26-5595
321
Jan. 21   20____
PAY TO THE ORDER OF Golden Valley Symphony Orchestra $ 50.00
Fifty and - - - - - - - - - - - - - - NO/100 DOLLARS
PALINDROME NATIONAL BANK
One Bankers Lane
Grand Island, MN 55553   (Signature)
                                                SIGNATURE
⑈001516 ⑆321080880⑆ 004189048409

Roscoe or Vivian Van Dusen
2331 Grapevine Lane
Grand View, MN 55551
(218) 555-1226
MDL# 358-DL-1298
383
26-5595
321
Jan. 25   20____
PAY TO THE ORDER OF Palindrome National Bank $ 75.00
Seventy-five and - - - - - - - - - - - NO/100 DOLLARS
PALINDROME NATIONAL BANK
One Bankers Lane
Grand Island, MN 55553   (Signature)
                                                SIGNATURE
⑈001516 ⑆321080880⑆ 004189048409

Student Page 48

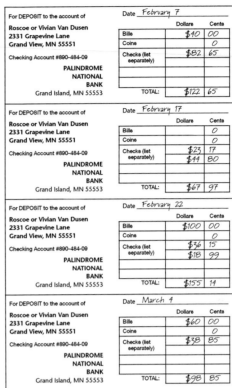

Roscoe or Vivian Van Dusen
2331 Grapevine Lane
Grand View, MN 55551
(218) 555-1226
MDL# 35B-DL-1298

384
26-5595
321

Jan. 25   20 __

PAY TO THE ORDER OF _Hearwell Telephone Co._ $ 100.00

One hundred and — — — — — — — — — NO/100 DOLLARS

PALINDROME NATIONAL BANK
One Bankers Lane
Grand Island, MN 55553

(Signature)
SIGNATURE

⑆001516 ⑆321080860⑆ 004189046409

Roscoe or Vivian Van Dusen
2331 Grapevine Lane
Grand View, MN 55551
(218) 555-1226
MDL# 35B-DL-1298

385
26-5595
321

Jan. 25   20 __

PAY TO THE ORDER OF _Bank of the North_ $ 75.00

Seventy-five and — — — — — — — — — NO/100 DOLLARS

PALINDROME NATIONAL BANK
One Bankers Lane
Grand Island, MN 55553

(Signature)
SIGNATURE

⑆001516 ⑆321080860⑆ 004189046409

Student Page 51

For DEPOSIT to the account of		Date February 7	
Roscoe or Vivian Van Dusen		Dollars	Cents
2331 Grapevine Lane	Bills	$40	00
Grand View, MN 55551	Coins		0
Checking Account #890-484-09	Checks (list separately)	$82	65
PALINDROME			
NATIONAL			
BANK			
Grand Island, MN 55553	TOTAL:	$122	65

For DEPOSIT to the account of		Date February 17	
Roscoe or Vivian Van Dusen		Dollars	Cents
2331 Grapevine Lane	Bills		0
Grand View, MN 55551	Coins		0
Checking Account #890-484-09	Checks (list separately)	$23	17
		$44	80
PALINDROME			
NATIONAL			
BANK			
Grand Island, MN 55553	TOTAL:	$67	97

For DEPOSIT to the account of		Date February 22	
Roscoe or Vivian Van Dusen		Dollars	Cents
2331 Grapevine Lane	Bills	$100	00
Grand View, MN 55551	Coins		0
Checking Account #890-484-09	Checks (list separately)	$36	15
		$18	99
PALINDROME			
NATIONAL			
BANK			
Grand Island, MN 55553	TOTAL:	$155	14

For DEPOSIT to the account of		Date March 4	
Roscoe or Vivian Van Dusen		Dollars	Cents
2331 Grapevine Lane	Bills	$60	00
Grand View, MN 55551	Coins		0
Checking Account #890-484-09	Checks (list separately)	$38	85
PALINDROME			
NATIONAL			
BANK			
Grand Island, MN 55553	TOTAL:	$98	85

Student Page 50

For DEPOSIT to the account of		Date January 3	
Roscoe or Vivian Van Dusen		Dollars	Cents
2331 Grapevine Lane	Bills	$30	00
Grand View, MN 55551	Coins		0
Checking Account #890-484-09	Checks (list separately)	$156	78
PALINDROME			
NATIONAL			
BANK			
Grand Island, MN 55553	TOTAL:	$186	78

For DEPOSIT to the account of		Date January 10	
Roscoe or Vivian Van Dusen		Dollars	Cents
2331 Grapevine Lane	Bills	$150	00
Grand View, MN 55551	Coins		0
Checking Account #890-484-09	Checks (list separately)	$650	00
PALINDROME			
NATIONAL			
BANK			
Grand Island, MN 55553	TOTAL:	$800	00

For DEPOSIT to the account of		Date January 18	
Roscoe or Vivian Van Dusen		Dollars	Cents
2331 Grapevine Lane	Bills	$30	00
Grand View, MN 55551	Coins	$20	00
Checking Account #890-484-09	Checks (list separately)	$150	00
		$250	00
PALINDROME		$50	00
NATIONAL			
BANK			
Grand Island, MN 55553	TOTAL:	$500	00

For DEPOSIT to the account of		Date January 24	
Roscoe or Vivian Van Dusen		Dollars	Cents
2331 Grapevine Lane	Bills		0
Grand View, MN 55551	Coins		0
Checking Account #890-484-09	Checks (list separately)	$75	00
		$75	00
PALINDROME		$25	00
NATIONAL			
BANK			
Grand Island, MN 55553	TOTAL:	$175	00

Student Page 53

RECORD ALL CHANGES OR CREDITS THAT AFFECT YOUR ACCOUNT

NUMBER	DATE	DESCRIPTION OF TRANSACTION	PAYMENT/DEBIT (-)		√ T	FEE (IF ANY) (-)	DEPOSIT/CREDIT (+)		BALANCE	
									173	63
372	1/6	Quad States Gas & Electric	43	40					130	23
373	1/6	A.J. Franklin Test.	25	00					105	23
—	1/10	Deposit					800	00	905	23
374	1/11	Comm. to Improve Our Schools	50	00					855	23
375	1/12	Pappas' Grocery	123	30					731	93
376	1/12	Best & Fitt Clothiers	85	00					646	93
377	1/15	Dr. I. Makum Welle	62	50					584	43
—	1/18	Deposit					500	00	1084	43
378	1/18	Cross-Country Airlines	232	84					851	59
379	1/18	Popp-Up Seed Company	18	50					833	09
380	1/21	Emma C. Payne, D.V.M.	35	40					797	69
381	1/21	NACUCOPAT	50	00					747	69
382	1/21	Golden Valley Symph. Orch.	50	00					697	69
—	1/24	Deposit					175	00	872	69
383	1/25	Palindrome National Bank	75	00					797	69
384	1/25	Hearwell Telephone Co.	100	00					697	69
385	1/25	Bank of the North	75	00					622	69

Student Page 54

RECORD ALL CHANGES OR CREDITS THAT AFFECT YOUR ACCOUNT

NUMBER	DATE	DESCRIPTION OF TRANSACTION	PAYMENT/DEBIT (-)		√ T	FEE (IF ANY) (-)	DEPOSIT/CREDIT (+)		BALANCE	
									622	69
386	2/1	Estevan Gutierrez	11	23					611	46
387	2/1	Northwest Cellular Phone	107	17					504	29
388	2/1	La Plaza	42	11					462	18
389	2/2	Estevan Gutierrez	13	89					448	29
390	2/2	Emma C. Payne, D.V.M.	25	50					422	79
391	2/2	Pappas' Grocery	41	20					381	59
392	2/4	Arnie Goldblum	8	95					372	64
—	2/7	Deposit					122	65	495	29
—	2/11	Deposit					808	35	1303	64
—	2/17	Deposit					67	97	1371	61
393	2/18	Super Charge Card	107	25					1264	36
—	2/22	Deposit					155	14	1419	50
394	2/24	Scanner Cable Television	60	00					1359	50
395	2/24	Hearwell Telephone Co.	58	25					1301	25
396	2/24	Quad States Gas & Electric	152	35					1148	90
—	2/25	Deposit					832	57	1981	47

Student Page 55

RECORD ALL CHANGES OR CREDITS THAT AFFECT YOUR ACCOUNT

NUMBER	DATE	DESCRIPTION OF TRANSACTION	PAYMENT/DEBIT (-)		√ T	FEE (IF ANY) (-)	DEPOSIT/CREDIT (+)		BALANCE	
									1981	47
397	3/1	Northwest Cellular Phones	42	18					1939	29
398	3/1	Estevan Gutierrez	17	25					1922	04
399	3/1	Quad States Gas & Electric	262	95					1659	09
—	3/4	Deposit					98	85	1757	94
400	3/7	Palindrome National Bank	1,022	85					735	09
401	3/7	Arthur C. Woo, M.D.	39	79					695	30
402	3/7	Super Charge Card	365	50					329	80
—	3/11	Deposit					808	35	1138	15
403	3/14	Rusty Bolts Service & Repair	61	00					1077	15
404	3/14	Estevan Gutierrez	32	50					1044	65
405	3/14	Bi-Well Market	9	85					1034	80
406	3/20	Bank of the North	213	79					821	01
407	3/21	Hearwell Telephone Co.	30	89					790	12
—	3/25	Deposit					832	57	1622	69
—	3/26	Deposit					250	00	1872	69
408	3/26	Mrs. Cynthia Broadwell	17	70					1854	99
409	3/26	Scanner Cable Television	25	65					1829	34

Student Page 56

RECORD ALL CHANGES OR CREDITS THAT AFFECT YOUR ACCOUNT

NUMBER	DATE	DESCRIPTION OF TRANSACTION	PAYMENT/DEBIT (-)		√ T	FEE (IF ANY) (-)	DEPOSIT/CREDIT (+)		BALANCE	
									1829	34
410	4/6	Pappas' Grocery	12	80					1816	54
411	4/6	Bi-Well Market	11	95					1804	59
412	4/6	Northwest Cellular	28	90					1775	69
413	4/6	Palindrome National Bank	1,973	37					-197	68
—	4/12	Deposit					808	35	610	67
—	4/13	Deposit					357	50	968	17
414	4/20	Super Charge Card	31	22					936	95
415	4/20	Estevan Gutierrez	11	00					925	95
416	4/20	PAWS auction	25	00					900	95
417	4/20	Hearwell Telephone Co.	32	95					868	00
418	4/20	Golden Valley PTA	50	00					818	00
—	4/23	Deposit					255	55	1073	55
419	4/24	UM Alumni Club	50	00					1023	55
420	4/25	Quad States Gas & Electric	185	50					838	05
421	4/26	Bank of the North	239	20					598	85
—	4/26	Deposit					832	57	1431	42
422	4/27	Scanner Cable Television	16	62					1414	80
—	4/27	Deposit					143	95	1558	75
423	4/28	Armando Rojas	14	07					1544	68
424	4/28	Larry's Drain Service	30	00					1514	68

Student Page 58

Front

STATEMENT OF ACCOUNT ACTIVITY

PALINDROME NATIONAL BANK
One Bankers Lane
Grand Island, MN 55553
Telephone (218) 555-3000

Roscoe or Vivian Van Dusen
2331 Grapevine Lane
Grand View, MN 55551

STATEMENT DATE: February 26, 2002

ACCOUNT NO: 890-484-09

DEPOSIT SUMMARY

DATE	AMOUNT/DESCRIPTION	DATE	AMOUNT/DESCRIPTION
01/04	$186.78		
01/10	$800.00		
01/18	$500.00		
01/24	$175.00		

CHECK SUMMARY

CHECK NO.	DATE	AMOUNT	CHECK NO.	DATE	AMOUNT
368	01/15	$35.00	376	01/29	$85.00
369	01/18	$260.00	377	01/29	$62.50
370	01/21	$96.70	379	01/28	$18.50
372	01/21	$43.40	382	01/28	$50.00
373	01/22	$25.00	384	02/03	$100.00
374	01/18	$50.00	385	02/03	$75.00

LAST STATEMENT BALANCE WAS	TO WHICH IS ADDED TOTAL DEPOSITS/CREDITS	AND SUBTRACTED TOTAL WITHDRAWALS/DEBITS	RESULTING IN A BALANCE OF
$514.00	$1661.78	$901.10	$1274.68

NOTICE: SEE REVERSE SIDE FOR IMPORTANT INFORMATION.

Student Page 59

Reverse Side

Use this form to reconcile the statement with your checkbook register.

Balancing Your Account

1. Place a check mark in your register next to any check that appears on this statement.

2. Place a check mark in your register next to any deposit that appears on this statement.

3. Place a check mark in your register next to any automatic deposit or electronic transfer that appears on this statement.

4. Any charges appearing on this statement but not appearing in your register should be deducted from your register balance before attempting to balance your register to this statement. Likewise, any credits appearing on this statement but not appearing in your register should be added to your register balance.

5. **ENDING BANK BALANCE** shown on this statement $ 1,274.68

6. **ADD (+) DEPOSITS** made but **not** shown on statement because made or received after date of this statement.

   TOTAL + $ 0

7. **SUBTRACT (–) TOTAL OF CHECKS OUTSTANDING** – $ 651.99 <<

8. **BALANCE** (Should agree with your adjusted register balance) = $ 622.69

CHECKS OUTSTANDING (Written but **not** shown on statement because not yet received by bank.)	
CHECK NUMBER	**AMOUNT**
371	135.45
375	123.30
378	232.84
380	35.40
381	50.00
383	75.00
TOTAL	651.99

Student Page 60

Front

STATEMENT OF ACCOUNT ACTIVITY

PALINDROME NATIONAL BANK
One Bankers Lane
Grand Island, MN 55553
Telephone (218) 555-3000

Roscoe or Vivian Van Dusen
2331 Grapevine Lane
Grand View, MN 55551

STATEMENT DATE: March 26, 2002

ACCOUNT NO: 890-484-09

DEPOSIT SUMMARY

DATE	AMOUNT/DESCRIPTION	DATE	AMOUNT/DESCRIPTION
02/07	$122.65	02/11	$808.35
02/17	$67.97	02/22	$155.14
02/25	$832.57		

CHECK SUMMARY

CHECK NO.	DATE	AMOUNT	CHECK NO.	DATE	AMOUNT
371	02/05	$135.45	386	02/12	$11.23
375	02/05	$123.30	387	02/12	$107.17
378	02/06	$232.84	388	02/13	$42.11
380	02/06	$35.40	389	02/19	$13.89
381	02/06	$50.00	393	02/20	$107.25
383	02/06	$75.00	394	03/02	$60.00

LAST STATEMENT BALANCE WAS	TO WHICH IS ADDED TOTAL DEPOSITS/CREDITS	AND SUBTRACTED TOTAL WITHDRAWALS/DEBITS	RESULTING IN A BALANCE OF
$1274.68	$1986.68	$993.64	$2267.72

NOTICE: SEE REVERSE SIDE FOR IMPORTANT INFORMATION.

Student Page 61

Reverse Side

Use this form to reconcile the statement with your checkbook register.

Balancing Your Account

1. Place a check mark in your register next to any check that appears on this statement.

2. Place a check mark in your register next to any deposit that appears on this statement.

3. Place a check mark in your register next to any automatic deposit or electronic transfer that appears on this statement.

4. Any charges appearing on this statement but not appearing in your register should be deducted from your register balance before attempting to balance your register to this statement. Likewise, any credits appearing on this statement but not appearing in your register should be added to your register balance.

5. **ENDING BANK BALANCE** shown on this statement $ 2,267.72

6. **ADD (+) DEPOSITS** made but **not** shown on statement because made or received after date of this statement.

   TOTAL + $ 0

7. **SUBTRACT (–) TOTAL OF CHECKS OUTSTANDING** – $ 286.25 <<

8. **BALANCE** (Should agree with your adjusted register balance) = $ 1,981.47

CHECKS OUTSTANDING (Written but **not** shown on statement because not yet received by bank.)	
CHECK NUMBER	**AMOUNT**
390	25.50
391	41.20
392	8.95
395	58.25
396	152.35
TOTAL	286.25

Student Page 62

Front

STATEMENT OF ACCOUNT ACTIVITY

PALINDROME NATIONAL BANK
One Bankers Lane
Grand Island, MN 55553
Telephone (218) 555-3000

Roscoe or Vivian Van Dusen
2331 Grapevine Lane
Grand View, MN 55551

STATEMENT DATE: April 26, 2002
ACCOUNT NO: 890-484-09

DEPOSIT SUMMARY			
DATE	AMOUNT/DESCRIPTION	DATE	AMOUNT/DESCRIPTION
03/04	$98.35	03/11	$808.35
03/25	$832.57	03/26	$250.00

CHECK SUMMARY					
CHECK NO.	DATE	AMOUNT	CHECK NO.	DATE	AMOUNT
390	03/05	$25.50	401	03/14	$39.79
391	03/05	$41.20	402	03/14	$365.50
395	03/05	$58.25	403	03/21	$61.00
396	03/07	$152.35	404	03/21	$32.50
397	03/12	$42.18	409	04/02	$25.65
398	03/12	$17.25			

LAST STATEMENT BALANCE WAS	TO WHICH IS ADDED TOTAL DEPOSITS/CREDITS	AND SUBTRACTED TOTAL WITHDRAWALS/DEBITS	RESULTING IN A BALANCE OF
$2267.72	$1989.27	$861.17	$3395.82

NOTICE: SEE REVERSE SIDE FOR IMPORTANT INFORMATION.

Student Page 64

Front

STATEMENT OF ACCOUNT ACTIVITY

PALINDROME NATIONAL BANK
One Bankers Lane
Grand Island, MN 55553
Telephone (218) 555-3000

Roscoe or Vivian Van Dusen
2331 Grapevine Lane
Grand View, MN 55551

STATEMENT DATE: May 26, 2002
ACCOUNT NO: 890-484-09

DEPOSIT SUMMARY			
DATE	AMOUNT/DESCRIPTION	DATE	AMOUNT/DESCRIPTION
04/12	$808.35	04/13	$357.50
04/23	$255.55	04/26	$832.57
04/27	$143.95		

CHECK SUMMARY					
CHECK NO.	DATE	AMOUNT	CHECK NO.	DATE	AMOUNT
392	04/05	$8.95	412	04/19	$28.90
400	04/05	$1,022.85	413	04/20	$1,973.37
405	04/05	$9.85	416	04/22	$25.00
406	04/05	$213.79	418	04/26	$50.00
407	04/08	$30.89	420	04/26	$185.50
408	04/08	$17.70	421	05/03	$239.20
410	04/19	$12.80	424	05/03	$30.00
411	04/19	$11.95			

LAST STATEMENT BALANCE WAS	TO WHICH IS ADDED TOTAL DEPOSITS/CREDITS	AND SUBTRACTED TOTAL WITHDRAWALS/DEBITS	RESULTING IN A BALANCE OF
$3395.82	$2397.92	$3860.75	$1932.99

NOTICE: SEE REVERSE SIDE FOR IMPORTANT INFORMATION.

Student Page 63

Reverse Side

Use this form to reconcile the statement with your checkbook register.

Balancing Your Account

1. Place a check mark in your register next to any check that appears on this statement.

2. Place a check mark in your register next to any deposit that appears on this statement.

3. Place a check mark in your register next to any automatic deposit or electronic transfer that appears on this statement.

4. Any charges appearing on this statement but not appearing in your register should be deducted from your register balance before attempting to balance your register to this statement. Likewise, any credits appearing on this statement but not appearing in your register should be added to your register balance.

5. **ENDING BANK BALANCE** shown on this statement $ 3,395.82

6. **ADD (+) DEPOSITS** made but **not** shown on statement because made or received after date of this statement.

    TOTAL + $ 0

7. **SUBTRACT (–) TOTAL OF CHECKS OUTSTANDING** – $ 1,566.98 <<

8. **BALANCE** (Should agree with your adjusted register balance) = $ 1,828.84

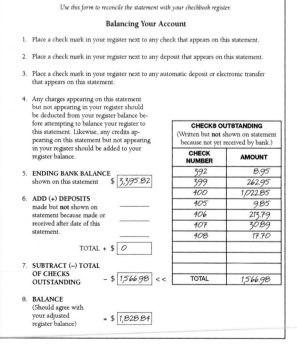

CHECKS OUTSTANDING (Written but **not** shown on statement because not yet received by bank.)	
CHECK NUMBER	AMOUNT
392	8.95
399	262.95
400	1,022.85
405	9.85
406	213.79
407	30.89
408	17.70
TOTAL	1,566.98

Student Page 65

Reverse Side

Use this form to reconcile the statement with your checkbook register.

Balancing Your Account

1. Place a check mark in your register next to any check that appears on this statement.

2. Place a check mark in your register next to any deposit that appears on this statement.

3. Place a check mark in your register next to any automatic deposit or electronic transfer that appears on this statement.

4. Any charges appearing on this statement but not appearing in your register should be deducted from your register balance before attempting to balance your register to this statement. Likewise, any credits appearing on this statement but not appearing in your register should be added to your register balance.

5. **ENDING BANK BALANCE** shown on this statement $ 1,932.99

6. **ADD (+) DEPOSITS** made but **not** shown on statement because made or received after date of this statement.

    TOTAL + $ 0

7. **SUBTRACT (–) TOTAL OF CHECKS OUTSTANDING** – $ 418.81 <<

8. **BALANCE** (Should agree with your adjusted register balance) = $ 1,514.18

CHECKS OUTSTANDING (Written but **not** shown on statement because not yet received by bank.)	
CHECK NUMBER	AMOUNT
399	262.95
414	31.22
415	11.00
417	32.95
419	50.00
422	16.62
423	14.07
TOTAL	418.81

Chapter 4

SAVINGS ACCOUNTS

Teaching Notes

Answers to Text Questions, page 68

1. The money received by a bank from an investor is used by the bank for making other investments. For example, the bank might purchase government bonds or lend money to people for mortgages. The income received by the bank is greater than the interest the bank pays the investor. For example, banks currently earn anywhere from 9% to 14% on the money that they lend to individuals and businesses. They pay about 1% to 3% on regular savings accounts. Thus, banks commonly earn a net profit of approximately 10% on these financial transactions.

2. The decision on how to invest money involves many factors. A regular savings account often has a low minimum-deposit requirement, such as $100. The bank pays interest on that account, as long as a minimum balance is maintained. The investor is allowed to withdraw money at any time. The interest rate on such accounts is usually very low.

   Money invested in insurance, on the other hand, is regarded as a long-term investment that pays relatively low interest because it provides death benefits as well as savings interest.

   Stocks are usually regarded as more risky investments because the return they provide may vary over time. Usually, no specific return is guaranteed on stocks and, as a result, one can realize significant gains or losses. Bonds are generally safer investments, with a certain standard return guaranteed to the bondholder by a governmental body or a corporation. Bond market values fluctuate, as do stock values, but usually over a narrower range. Bonds mature, as do insurance policies, although often in somewhat less time.

3. Saving money is almost a necessity in our society. Most families are confronted from time to time with fairly large expenses, such as the purchase of a new house or new car, or the cost of a college education, for which money should be set aside. Unexpected events, such as a major illness, may require large sums of money for which savings would be very important.

4. One purpose of a life-insurance policy is to provide financial protection to the family and friends of a policyholder in the event of his or her death. The insurance policy can also be thought of as a savings account, however. The insurance company pays interest to the policyholder on all the money that he or she has paid in premiums over the years. The amount of interest paid on an insurance policy, however, is usually less than that on a regular savings account since, in the former case, the policyholder is also receiving the benefits of life-insurance protection in case of death.

5. Suppose that a bank pays a certain amount of interest on money in a savings account. If that interest is then added to the account and included in the principal the next time interest is figured, we say that the interest is being *compounded*. If, on the other hand, that interest is not added to the account (it might be paid out directly, for example), then the interest paid is *simple* interest.

6. Credit unions are financial institutions organized to make loans to those who might not otherwise be eligible at a commercial bank. A credit union is a "democratic" organization in the sense that everyone who conducts business at the credit union must become a

member and then has a vote in its operation. This system does not operate in commercial banks. Many credit unions are established by companies, unions, or other groups of people who are in some way connected as, for example, a teachers' credit union.

7. A *certificate of deposit,* or CD, is a financial instrument that allows a person to deposit a sum of money at a savings institution for a period of time at a rate of interest that is usually higher than the rate for a regular savings account. Certificates of deposit may be issued for any period of time from a month to five years or more. In general, the longer the period of time for which the CD is issued, the higher the rate of interest it carries. A disadvantage of CDs is that the investor cannot withdraw the money without penalty before a specified period of time.

Student Page 69

STATEMENT OF ACCOUNT ACTIVITY

PALINDROME NATIONAL BANK
One Bankers Lane
Grand Island, MN 55553
Telephone (218) 555-3000

Roscoe or Vivian Van Dusen
2331 Grapevine Lane
Grand View, MN 55551

INTEREST RATE:
5.25% per year,
compounded monthly

ACCOUNT NO: 890-4456-4

SAVINGS ACCOUNT SUMMARY

DATE	DEPOSITS: AMOUNT/DESCRIPTION	WITHDRAWALS: AMOUNT/DESCRIPTION	BALANCE
Jan 1	Previous Balance		$289.90
Jan 15	200.00		489.90
Jan 30	2.14 INTEREST		492.04
Feb 8	175.00		667.04
Feb 19	45.00		712.04
Feb 26	50.00		762.04
Feb 26	3.33 INTEREST		765.37
Mar 5	25.00		790.37
Mar 23	325.00		1115.37
Mar 30	4.88 INTEREST		1120.25
Apr 5		225.00	895.25
Apr 23	100.00		995.25
Apr 30	4.35 INTEREST		999.60
May 5	100.00		1099.60
May 15	50.00		1149.60
May 23		43.50	1106.10
May 29	4.84 INTEREST		1110.94
June 3		863.44	247.50
June 11	250.00		497.50
June 18	300.15		797.65
June 25	156.47		954.12
June 29	4.17 INTEREST		958.29

REMINDER—BEGINNING JULY 1, YOUR INTEREST RATE CHANGES TO 5.55% PER YEAR, COMPOUNDED MONTHLY.

LAST STATEMENT BALANCE WAS	TO WHICH IS ADDED TOTAL DEPOSITS/CREDITS	AND SUBTRACTED TOTAL WITHDRAWALS/DEBITS	RESULTING IN A BALANCE OF
289.90	1,800.33	1,131.94	958.29

Student Page 70

STATEMENT OF ACCOUNT ACTIVITY

PALINDROME NATIONAL BANK
One Bankers Lane
Grand Island, MN 55553
Telephone (218) 555-3000

Roscoe or Vivian Van Dusen
2331 Grapevine Lane
Grand View, MN 55551

INTEREST RATE:
5.55% per year,
compounded monthly

ACCOUNT NO: 890-4456-4

SAVINGS ACCOUNT SUMMARY

DATE	DEPOSITS: AMOUNT/DESCRIPTION	WITHDRAWALS: AMOUNT/DESCRIPTION	BALANCE
July 1	Previous Balance		958.29
July 9	325.50		1283.79
July 15	78.60		1362.39
July 18		300.00	1062.39
July 21	193.22		1255.61
July 29	5.81 INTEREST		1261.42
Aug 1	150.00		1411.42
Aug 7		467.75	943.67
Aug 10	285.00		1228.67
Aug 20	91.45		1320.12
Aug 22	60.50		1380.62
Aug 29	6.39 INTEREST		1387.01
Sept 3		575.00	812.01
Sept 5	325.00		1137.01
Sept 14	185.85		1322.86
Sept 19	215.00		1537.86
Sept 23	175.00		1712.86
Sept 30	7.92 INTEREST		1720.78
Oct 10		625.00	1095.78
Oct 11	425.89		1521.67
Oct 15		135.47	1386.20
Oct 20	250.00		1636.20
Oct 25	316.95		1953.15
Oct 29	9.03 INTEREST		1962.18

REMINDER—BEGINNING JULY 1, YOUR INTEREST RATE CHANGED TO 5.55% PER YEAR, COMPOUNDED MONTHLY.

LAST STATEMENT BALANCE WAS	TO WHICH IS ADDED TOTAL DEPOSITS/CREDITS	AND SUBTRACTED TOTAL WITHDRAWALS/DEBITS	RESULTING IN A BALANCE OF
958.29	3,107.11	2,103.22	1,962.18

Student Page 72

MEMBER'S STATEMENT OF ACCOUNT

VLA CREDIT UNION
Golden Valley Unified District
Paradise, Minnesota 55511

Record of Account with: Vivian Van Dusen

Rules and Regulations of the Credit Union:

1. Shares cost $5 each.

2. Interest is $5\frac{3}{4}$% per annum, figures and compounded quarterly on full shares only.

SHARE ACCOUNT (NOT TRANSFERABLE)

DATE	PAID IN	WITHDRAWN	BALANCE	FULL SHARES
01/03	Balance Forward		$584.85	116
01/09	$150.00		$734.85	146
01/31	$250.00		$984.85	196
02/19	$125.00		$1109.85	221
02/26	$135.00		$1244.85	248
03/07	$200.00		$1444.85	288
03/15	$200.00		$1644.85	328
03/26		$250.00	$1394.85	278
03/30	$19.98 Interest		$1414.83	282
04/19	$283.50		$1698.33	339
05/05	$160.00		$1858.33	371
05/25		$349.50	$1508.83	301
06/05	$175.00		$1683.83	336
06/27	$24.15 Interest		$1707.98	341

Retain this statement as a part of your permanent records.

Student Page 74

MEMBER'S STATEMENT OF ACCOUNT

VLA CREDIT UNION
Golden Valley Unified District
Paradise, Minnesota 55511

Record of Account with: Vivian Van Dusen

Rules and Regulations of the Credit Union:

1. Shares cost $5 each.

2. Interest is $5\frac{3}{4}$% per annum, figures and compounded quarterly on full shares only.

SHARE ACCOUNT (NOT TRANSFERABLE)

DATE	PAID IN	WITHDRAWN	BALANCE	FULL SHARES
10/01	Balance Forward		$1183.74	236
10/03	$225.00		$1408.74	281
10/19	$275.50		$1684.24	336
10/22		$508.96	$1175.28	235
11/04	$150.00		$1325.28	265
11/16	$150.00		$1475.28	295
11/20	$87.75		$1563.03	312
11/30	$175.00		$1738.03	347
12/04	$165.67		$1903.70	380
12/08	$345.50		$2249.20	449
12/10		$1,958.87	$290.33	58
12/26	$575.68		$866.01	173
12/30	$12.43 Interest		$878.44	175

Retain this statement as a part of your permanent records.

Student Page 73

MEMBER'S STATEMENT OF ACCOUNT

VLA CREDIT UNION
Golden Valley Unified District
Paradise, Minnesota 55511

Record of Account with: Vivian Van Dusen

Rules and Regulations of the Credit Union:

1. Shares cost $5 each.

2. Interest is $5\frac{3}{4}$% per annum, figures and compounded quarterly on full shares only.

SHARE ACCOUNT (NOT TRANSFERABLE)

DATE	PAID IN	WITHDRAWN	BALANCE	FULL SHARES
07/01	Balance Forward		$1707.98	341
07/05	$150.00		$1857.98	371
07/18		$1,757.96	$100.02	20
07/28	$247.50		$347.52	69
08/06	$175.00		$522.52	104
08/17	$431.75		$954.27	190
08/23	$125.00		$1079.27	215
08/31	$168.50		$1247.77	249
09/01	$87.50		$1335.27	267
09/20		$275.00	$1060.27	212
09/24	$106.72		$1166.99	233
09/29	$16.75 Interest		$1183.74	236

Retain this statement as a part of your permanent records.

Comments on Internet Activities, pages 75–76

1. The balance at the end of one week in the account described in this question would be $500.50. Interest earned during the week would be $0.50.

   Some web sites explain the mathematical formula for calculating compound interest. That formula may be too complex for students using this book. Many web sites provide an automatic compound-interest calculator that can be used simply by inputting principal, interest rate, and time of investment. Students can use these calculators quite easily. Some web sites that students might find include the following:

   > Compound-Interest Calculations
   > http://www.warwick.ac.uk/fac/cross_fac/nflc/cal2.html
   >
   > Compound-Interest Calculator
   > http://www.1728.com/compint.htm
   >
   > Future Value JavaScript Calculator
   > http://www.webwinder.com/wwhtmbin/java_fv1.html

   The values of a $500 investment kept for one year at the interest rates given are as follows:

   > $5\frac{1}{4}$%: $526.89
   > $5\frac{1}{2}$%: $528.20
   > $5\frac{3}{4}$%: $529.52

2. Many Internet sites provide information on the Social Security system, but probably the best place to begin is the Social Security Administration's own web site at http://www.ssa.gov/. The retirement questions asked of students can be answered on the site's Quick Calculator page. The monthly benefits for each individual at ages 62, 66, and 70 as of publication are as follows:

   (a) $1,149; $1,535; $2,030

   (b) $957; $1,280; $1,695

   (c) $1,282; $1,837; $2,282

3. The most common retirement plans are traditional IRAs; Roth IRAs, Education IRAs; Simplified Employee Pensions (SEPs); Savings Incentive Match Plans for Employees (SIMPLE); and Keough plans. The details and provisions of each of these plans are complex. The best sources for information on the plans are IRS publications dealing with them, specifically, Publications 560 (Retirement Plans for Small Business) and 590 (Individual Retirement Plans). Both publications can be obtained by contacting the IRS on-line at the IRS web site, http://www.irs.gov/forms_pubs/.

   Information about a variety of retirement plans is available on-line from banks, financial advisors, retirement planners, and similar sources. Students may find other articles and publications that deal with this subject. One of the most useful sources is the Money Central web site at http://moneycentral.msn.com/, which has a variety of articles on various kinds of retirement plans.

Chapter 5

GROCERY SHOPPING

Teaching Notes

You may want to discuss with students the many factors that can affect the price of groceries to explain why the sample prices provided here may be very different from those obtained by your students.

Some of the factors that account for differences in food prices include: (1) variations found at different types of stores (discount, specialty, chain or convenience stores, for example); (2) variations found within one chain, depending on geographical location (urban, suburban or rural, for example); (3) variations among brands (national, regional or generic brands, for example); (4) variations owing to special sales; and (5) seasonal variations (for fresh fruits and vegetables at various times of the year).

You may also want to discuss the difficulties comparing prices when package sizes differ. For example, some foods are packaged in one-pound or one-quart containers. Others may be packaged in fractional weight or liquid measures. Canned vegetables might be available in $14\frac{1}{2}$-ounce sizes. The size of such containers is largely a matter of tradition and competition.

Because of the variation in package sizes, there are a number of options on how to complete the charts provided in the Activity Text. Some students may round off fractional weights and volumes to the next nearest whole number before making calculations. Thus, a $14\frac{1}{2}$-ounce package could be called 14 or 15 ounces for the purpose of calculations of unit price. Others may do calculations using the exact weight or volume of a package. In that way, you can give students practice in mathematical operations that involve fractions and decimals. The unit price for a can of green beans that weighs $14\frac{1}{2}$ ounces and costs \$1.39, for example, is found by dividing a decimal (\$1.39) by a mixed number ($14\frac{1}{2}$ oz.) transformed into a decimal (14.5).

Answers to Text Questions, page 78

1. Similar items sold by different food producers are often available in various and odd sizes. For example, tuna fish is sold in $3\frac{1}{2}$-ounce, $7\frac{1}{4}$-ounce, and 14-ounce sizes. *Unit pricing* tells a shopper how much each item costs in some standard measurement, such as price per ounce, price per pound, or price per gallon. This practice allows the shopper to compare the price of a product in any size package.

2. Stores sometimes use sales to lure customers into a store. However, there may be only a limited amount of the sale item on hand, or items may not be in good condition. And the price of items not on sale may be so high that the customer actually spends more at the store, rather than saving money. Still another factor to consider is the original price of the sale item. It could be that the original price was so high that, even on sale, the customer realizes no savings in comparison with prices at other stores.

3. *Convenience foods* are packaged foods that may be partially cooked so that minimum preparation time is needed. Canned vegetables and frozen dinners are examples of convenience foods. Although convenience foods can save time and effort in preparing a meal, they have certain drawbacks. They are usually more expensive than the price of raw materials. They generally contain food additives that may not be nutritious and, in some cases, may actually be harmful. Finally, they are seldom as enjoyable to eat as a meal prepared from scratch.

4. Stores can make extensive use of advertising on television and radio and in newspapers to attract buyers. They can also construct attractive displays in the store to catch the buyer's eye. Special sales and the offer of coupons for free or reduced-price items are also available.

   Some stores and chains now have food clubs that consumers can join. Club members receive special discounts on items available in the store or from the chain. It should be noted that stores often emphasize luxury or impulse items not necessary to a healthy diet but that a customer may be talked into buying. The placement of such items at the front of the store, at the end of an aisle, or in some other conspicuous location may increase the chance that a consumer will buy the items.

5. During some seasons of the year, packaged items may be less expensive than unprocessed foods because of the difficulty and expense of obtaining such raw foods. For example, during the winter, canned pineapple is likely to be less expensive than fresh pineapple.

Comments on Internet Activities, page 79

1. The Food Stamp Program is an activity funded by federal or state governments by which certain individuals can receive food stamps. These can be used to buy food at most groceries. The Food Stamp Program is designed for low-income individuals and families. Eligibility for food stamps is based on a number of factors, primarily household income.

   The Electronic Benefit Transfer program allows the transfer of the dollar value of food stamps directly to a specific retailer. For example, rather than sending $100 worth of food stamps to a person or family, the government sends a voucher worth $100 to a store of the recipient's choice.

   The best source of detailed information about the Food Stamp Program is the U.S. Department of Agriculture's web site at http://www.fns.usda.gov/fsp/. The Electronic Benefit Transfer program is described in a separate section of that site at http://www.fns.usda.gov/fsp/menu/admin/ebt/ebt.htm.

2. Information about smart shopping skills can be found on a number of web sites devoted to consumer information. Some sites focus specifically on how to teach children to become intelligent shoppers. Two articles of this type can be found at the respective sites listed below:

   > eHow to Take Children Grocery Shopping
   > http://www.ehow.com/

   > Grocery Shopping Skills for Kids
   > http://financialfinesse.com/

3. The U.S. Department of Agriculture has an extensive list of publications on meal planning. Similar information is available from many state and public-interest agencies. These publications are usually available in hard copy as well as on the Internet. Examples of some Internet sources that may be of value in meal planning include the following:

   > Guide to Menu Planning
   > http://www.cdasandiego.com/Menu.htm

   > Meal Planning (a collection of related web sites)
   > http://www.busycooks.about.com/

   > Meal Planning and Make-Ahead Cooking—The Kitchen Link
   > http://www.kitchenlink.com/rcpmenus.html

4. Internet grocery shopping is largely a regional activity because of the problem of making deliveries from a store to nearby customers. Such programs have become popular in some parts of the country but less so in other regions. One criterion for success is a population density that would offset delivery costs.

Some stores and chains that currently provide
Internet grocery-shopping sites are the
following:

Groceries to Go
http://www.grocerywagon.com/

Peapod
http://www.peapod.com/

Answers will vary.

Student Page 81

Worksheet for Advertising and Food Buying

The following list suggests a few items that you might check at the stores you visit. Add as many items as you like to complete this list. Be sure to note the brand name and the size of the item for which you get a price.

Item	Quantity	Price	Store 1	Store 2	Store 3
vegetable oil	48 oz	advertised	$ 2.39	$ 2.49	$ 2.69
		regular	$ 2.99	$ 2.99	$ 3.09
		price difference	$ 0.60	$ 0.50	$ 0.40
		percentage difference	20 %	17 %	13 %
fresh potatoes	5 lb bag	advertised	$ 2.49	$ 2.39	$ 1.99
		regular	$ 2.99	$ 2.99	$ 2.39
		price difference	$ 0.50	$ 0.60	$ 0.40
		percentage difference	17 %	22 %	17 %
ground beef	1 lb	advertised	$ 1.19	$ 1.19	$ 1.19
		regular	$ 1.39	$ 1.99	$ 1.59
		price difference	$ 0.20	$ 0.80	$ 0.40
		percentage difference	14 %	40 %	25 %
strawberry jelly	1 lb	advertised	$ 3.29	$ 3.19	$ 3.59
		regular	$ 3.49	$ 3.49	$ 3.89
		price difference	$ 0.20	$ 0.30	$ 0.30
		percentage difference	6 %	9 %	8 %

Answers will vary.

Student Page 83

Item	Amount	Unprocessed (raw)	Prepared	Price Difference	Percentage Difference
Potatoes	1 lb	$ 0.39	$ 2.94	$ 2.55	+654 %
Peas	1 lb	$ 1.98	$ 0.99	$ 0.99	−50 %
Orange juice	1 qt	$ 2.59	$ 1.99	$ 0.60	−23 %
Peanuts	1 lb	$ 1.99	$ 3.59	$ 1.60	+80 %
Milk	1 qt	$ 1.35	$ 1.32	$ 0.03	−2 %
Corn	1 lb	$ 1.25	$ 1.59	$ 0.34	+27 %
		$	$	$	%
		$	$	$	%
		$	$	$	%
		$	$	$	%
		$	$	$	%
		$	$	$	%
		$	$	$	%
		$	$	$	%
		$	$	$	%
		$	$	$	%
		$	$	$	%
		$	$	$	%
		$	$	$	%
		$	$	$	%
		$	$	$	%
		$	$	$	%
		$	$	$	%

Answers will vary.

Student Page 85

We have started you with a few items that come in large, medium, and small packages. When you visit the grocery store, add any other items that you can find that also come in packages of more than one size.

ITEM	SMALL PACKAGE			MEDIUM PACKAGE			LARGE PACKAGE		
	Size	Price	Unit Price	Size	Price	Unit Price	Size	Price	Unit Price
Sugar	1 lb.	$0.89	$0.89	5 lb.	$3.99	$0.80	25 lb.	$9.99	$0.40
Peanuts	12 oz.	$3.49	$4.65	16 oz.	$3.59	$3.59	22 lb.	$7.19	$3.26
Dry Milk	9.6 oz.	$1.49	$2.48	25.6 oz.	$3.69	$2.31	4 lb.	$7.29	$1.82
Apple Juice	10 oz.	$0.59	$0.94	46 oz.	$2.49	$0.87	128 oz.	$4.99	$0.62
Peanut Butter	12 oz.	$1.79	$2.39	16 oz.	$2.49	$2.49	4 lb.	$7.19	$1.80
Vegetable Oil	32 oz.	$2.69	$1.34	48 oz.	$2.99	$1.00	64 oz.	$3.79	$0.95

Chapter 6

EXPENSES AROUND THE HOUSE

Teaching Notes

Answers to Text Questions, pages 87–88

1. A general store is usually a local store that carries a limited selection of commonly used products. Although convenient for last-minute purchases of small items, such stores are likely to be more expensive than larger stores. In rural areas, a general store may be the only one available.

   A department store generally carries a larger variety of items, usually at lower cost than smaller neighborhood stores. Specialty stores are likely to offer a large variety of items within a specific category of product. A button shop, for example, may carry hundreds or thousands of types of buttons; a department store may have only a few dozen kinds.

   Discount stores carry essentially the same items that a food store, a department store, or a hardware store might carry, but their prices tend to be lower. Lower prices at discount stores are usually possible because of less elaborate interiors and fewer salespeople.

   Mail-order houses probably offer the largest variety of items to be found anywhere. Prices may be lower than those of department stores, but you must wait before receiving your purchase; you may have trouble having your product serviced; and you cannot always examine the item before purchasing it. In addition, you have to pay the cost of shipping the product to you.

   Shopping on the Internet has many of the advantages and disadvantages of mail-order shopping, except that it is quicker and easier to order on-line than to fill out a form for a mail-order purchase.

2. Discount stores' profits depend on selling a large number of items in simple surroundings, usually with fewer salespeople.

3. A credit card enables you to make purchases without having to carry a lot of cash. You may want to make a purchase when funds are not immediately available. The credit card is helpful because you can purchase the item and pay for it later. In many cases, you can also borrow money or withdraw money from an ATM with a credit card. Credit card charge slips provide simple receipts for tax-deductible expenses or for any items for which you may want a permanent record.

   On the other hand, items not paid for upon receipt of the credit card bill begin to accrue interest charges. In such cases, purchases end up costing more when charged than if they had been paid for with cash. Charging with a credit card is so easy that many people spend more with a credit card than they might if they paid with cash or by check. Most financial experts believe that many Americans carry a much higher credit-card debt than is wise or safe.

4. Taking financial advantage of customers is hardly a new practice in the retail business; a complete list of techniques is virtually endless. Some of the more obvious practices include figuring the sales tax incorrectly; giving the wrong change; adding the bill incorrectly; and figuring a discount incorrectly.

   Customers can avoid being cheated by paying attention when a sale is made. The customer should recheck the merchant's calculations to ensure there is no error, not even an honest one. A merchant suspected of cheating can be taken to small-claims court. Reports should also be made to the local Better Business Bureau. Many states and communities have consumer advocates to whom cheating should be reported.

Having said all this, it is important to point out that financial dishonesty in retail stores has probably dropped dramatically in the last half century. One reason for this improvement is that most transactions are now carried out electronically, so that the chance of human error or deception is much reduced. Consumers have also probably become better informed about redress.

Comments on Internet Activities, page 88

1. It is now possible to apply for a number of credit cards on-line. In fact, some credit cards have been created as an Internet product; Next Card is one example. Some credit card companies have taken to advertising and soliciting aggressively on-line. The Capitol Bank is one such provider.

    The information required to apply for a credit card can range from minimal to extensive. In addition to name, address, and telephone number, applicants may also be asked to provide their Social Security number and annual personal or household income. They may be asked to provide financial references and other personal information.

2. Some of the many varieties of credit cards available are bank credit cards, automobile credit cards, student credit cards, Internet credit cards, grocery and retail store credit cards, frequent-flyer credit cards, business and corporate credit cards, and cash-back credit cards.

    Credit cards differ from each other in some fundamental ways. They may or may not have annual fees. The grace period, during which no payment is necessary, varies, as does the finance charge (interest rate) and the way it is determined. Credit cards may or may not provide a cash-back bonus at the end of the year. Dollars spent with the card may or may not be redeemable as mileage on an airline; with a hotel chain or car rental company; or with some other retail business.

3. Information on consumer protection agencies is easily located through a web search of the term *consumer protection*. The web site for the U.S. Federal Trade Commission Office of Consumer Protection is located at http://www.ftc.gov/ftc/consumer.htm/. The web site for the U.S. Consumer Protection Safety Commission is at http://www. cpsc.gov/. A search for *consumer protection* is also likely to yield web pages for state consumer protection agencies and for agencies and organizations with interest in special areas, such as car purchases and consumer protection for the blind.

Example — Student Page 89

ADVERTISED ITEM	STORE WHERE ADVERTISED Name & Price	STORE 2 Name & Price	STORE 3 Name & Price	LARGEST PRICE DIFFERENCE FOUND	PERCENTAGE DIFFERENCE
Recliner chair	Salem Furniture $399.95	Orono's Furniture $329.95	Sam's Discount $298.95	$101.00	33.8%
Couch	Salem Furniture $499.99	Orono's Furniture $429.95	Sam's Discount $399.95	$100.04	25%
Table lamp	Salem Furniture $35.95	Orono's Furniture $55.95	Sam's Discount $38.99	$20.00	55.6%
Floor lamp	Salem Furniture $79.95	Orono's Furniture $109.99	Sam's Discount $69.95	$40.04	57.2%
End table	Salem Furniture $134.50	Orono's Furniture $185.95	Sam's Discount $117.99	$67.96	57.6%

Answers will vary

Student Page 90

Exercise II. Deciding Where to Shop

Deciding where to buy an item can depend on many factors. Price, of course, is important. So is service. What other factors might influence your choice of stores?

In this exercise, you will concentrate on only one of these factors: cost. Listed below are a number of purchases the Van Dusens want to make over the next few months. Space is provided for you to fill in other items you might want to price. Record the cost of each item on your list at each store you visit. One column is provided for any Internet price that you can find. Finally, calculate the price differences and percentage differences between the highest and lowest prices for each item on your list.

Item	Store 1 Price	Store 2 Price	Store 3 Price	Internet Price	Largest Price Difference	Percentage Difference
Blue denim workshirt, size large	$17.95	$21.50	$15.99	$14.99	$6.51	43.4%
Cold cream, 1-lb jar	$3.99	$5.19	$3.17	- - -	$2.02	63.7%
9 × 12 braided rug for sunroom	$109.79	$155.00	$99.99	$109.95	$55.01	35%
First baseman's baseball mitt	$37.50	$45.00	$32.99	$42.79	$12.01	36.4%
Best-selling CD of the week	$14.98	$14.98	$12.79	$14.98	$2.19	17.1%
Child's summer dress, size 14	$24.79	$31.00	$16.95	$20.35	$14.05	82.9%

Student Page 92

The SUPER CHARGE CARD

D046025

Account Number	Statement Date
8322-156-00938	February 1

Roscoe and Vivian Van Dusen
2331 Grapevine Lane
Grand View, MN 55551

The Super Charge Card
Happy Valley State Bank
1425 W. Woodrow Wilson Blvd.
Montrose, TX 78712

Please indicate any changes below:
Name _____
Address _____
City, State _____ Zip Code ____
Telephone _____

Please be sure our mailing address above appears in the window of the return envelope.

Please include your account number on your check and on any correspondence.

Return this stub with your check.

Detach payment stub here.   Amount paid _____

Transaction Date	Posting Date	Reference	Merchant Name or Transaction Description	New Purchases, Fees, Advances & Debits	Payments & Credits
1/03	1/06	498876739938-2	Town Paint and Supply	$42.55	
1/05	1/10	498839827481-1	Zippo Gasoline Company	$8.92	
	1/15		Payment, Thank You		$50.00
1/15	1/20	493882901982-6	Hardeetax Lumber Company	$89.96	
1/21	1/25	493001546872-7	Kitchen DeeLight Reastarant	$33.89	
1/28	1/31	493023870361-7	Bardelow's Department Store	$25.67	

Previous Balance	New Purchases, Fees, Advances & Debits	Finance Charge (Due to Periodic Rate)	Payments & Credits	New Balance	Minimum Payment Due
$324.80	$200.99	$4.12	$50.00	$479.91	$24.00

Monthly Periodic Rate	Corresponding Annual Percentage Rate	Balances to Which Applicable	FOR OFFICE USE ONLY
1.5%	18.00%	PURCHASES, ADVANCES, FINANCE CHARGES & FEES	

Student Page 93

The SUPER CHARGE CARD

D046025

Account Number	Statement Date
8322-156-00938	March 1

Roscoe and Vivian Van Dusen
2331 Grapevine Lane
Grand View, MN 55551

The Super Charge Card
Happy Valley State Bank
1425 W. Woodrow Wilson Blvd.
Montrose, TX 78712

Please indicate any changes below:
Name _____
Address _____
City, State _____ Zip Code ____
Telephone _____

Please be sure our mailing address above appears in the window of the return envelope.

Please include your account number on your check and on any correspondence.

Return this stub with your check.

Detach payment stub here.   Amount paid _____

Transaction Date	Posting Date	Reference	Merchant Name or Transaction Description	New Purchases, Fees, Advances & Debits	Payments & Credits
	2/06		Payment, Thank You		$125.00
2/06	2/09	498203928182-2	Custom-eeze	$35.89	
2/14	2/17	482938103947-6	The Old Shoppe	$55.90	
2/19	2/24	498203987342-6	Speedmobile Repairs	$139.95	
2/20	2/24	492839280240-2	Cromwell Books	$18.75	
2/20	2/25	487928739843-3	Dandy Candy	$12.69	
2/24	2/27	490928983343-5	Dr. Carlton	$69.00	
2/26	2/29	489988812032-4	Dr. Carlton	$22.50	

Previous Balance	New Purchases, Fees, Advances & Debits	Finance Charge (Due to Periodic Rate)	Payments & Credits	New Balance	Minimum Payment Due
$479.91	$354.48	$5.32	$125.00	$714.71	$35.74

Monthly Periodic Rate	Corresponding Annual Percentage Rate	Balances to Which Applicable	FOR OFFICE USE ONLY
1.5%	18.00%	PURCHASES, ADVANCES, FINANCE CHARGES & FEES	

Student Page 94

The SUPER CHARGE CARD
D046025

Account Number	Statement Date
8322-156-00938	April 1

Roscoe and Vivian Van Dusen
2331 Grapevine Lane
Grand View, MN 55551

The Super Charge Card
Happy Valley State Bank
1425 W. Woodrow Wilson Blvd.
Montrose, TX 78712

Please indicate any changes below:
Name _____
Address _____
City, State _____ Zip Code _____
Telephone _____

Please be sure **our** mailing address above appears in the window of the return envelope.

Detach payment stub here.

Please include your account number on your check and on any correspondence.

Return this stub with your check.

Amount paid _____

Transaction Date	Posting Date	Reference	Merchant Name or Transaction Description	New Purchases, Fees, Advances & Debits	Payments & Credits
3/02	3/05	498838476928-9	Zippo Gasoline Company	$19.58	
3/02	3/08	498900988278-0	Kitchen DeeLight Reataurant	$53.42	
3/04	3/13	498837780982-0	The Trap	$163.44	
	3/17		Payment, Thank You		$175.00
3/14	3/18	498834447091-2	Kitchen DeeLight Reataurant	$41.42	
3/14	3/18	498839870928-3	Muller's	$193.47	
3/14	3/21	498930998978-2	The Daily Plant	$38.83	
3/20	3/23	498900989093-1	Zippo Gasoline Company	$59.97	
3/21	3/25	498899993132-6	Kitchen DeeLight Reataurant	$19.89	

Previous Balance	New Purchases, Fees, Advances & Debits	Finance Charge (Due to Periodic Rate)	Payments & Credits	New Balance	Minimum Payment Due
$714.71	$590.02	$8.10	$175.00	$1137.83	$56.89

Monthly Periodic Rate	Corresponding Annual Percentage Rate	Balances to Which Applicable	FOR OFFICE USE ONLY
1.5%	18.00%	PURCHASES, ADVANCES, FINANCE CHARGES & FEES	

Student Page 95

The SUPER CHARGE CARD
D046025

Account Number	Statement Date
8322-156-00938	May 1

Roscoe and Vivian Van Dusen
2331 Grapevine Lane
Grand View, MN 55551

The Super Charge Card
Happy Valley State Bank
1425 W. Woodrow Wilson Blvd.
Montrose, TX 78712

Please indicate any changes below:
Name _____
Address _____
City, State _____ Zip Code _____
Telephone _____

Please be sure **our** mailing address above appears in the window of the return envelope.

Detach payment stub here.

Please include your account number on your check and on any correspondence.

Return this stub with your check.

Amount paid _____

Transaction Date	Posting Date	Reference	Merchant Name or Transaction Description	New Purchases, Fees, Advances & Debits	Payments & Credits
4/01	4/10	503872093212-9	Scripps Bookstore	$15.68	
	4/12		Payment, Thank You		$100.00
4/01	4/14	504983928902-4	Hammer and Nale Hardware	$22.40	
4/14	4/16	505988498392-6	Save-More Gasoline	$8.52	
4/20	4/23	505982883894-5	Taetee Time Reataurant	$18.95	
4/20	4/27	505928398374-1	The Bent Toe Shoe Shoppe	$26.78	
4/25	4/28	504872983474-3	HoneyDo Bakery	$15.50	

Previous Balance	New Purchases, Fees, Advances & Debits	Finance Charge (Due to Periodic Rate)	Payments & Credits	New Balance	Minimum Payment Due
$1137.83	$107.83	$15.57	$100.00	$1161.23	$58.06

Monthly Periodic Rate	Corresponding Annual Percentage Rate	Balances to Which Applicable	FOR OFFICE USE ONLY
1.5%	18.00%	PURCHASES, ADVANCES, FINANCE CHARGES & FEES	

Student Page 96

The SUPER CHARGE CARD
D046025

Account Number	Statement Date
8322-156-00938	June 1

Roscoe and Vivian Van Dusen
2331 Grapevine Lane
Grand View, MN 55551

The Super Charge Card
Happy Valley State Bank
1425 W. Woodrow Wilson Blvd.
Montrose, TX 78712

Please indicate any changes below:
Name _____
Address _____
City, State _____ Zip Code _____
Telephone _____

Please be sure **our** mailing address above appears in the window of the return envelope.

Detach payment stub here.

Please include your account number on your check and on any correspondence.

Return this stub with your check.

Amount paid _____

Transaction Date	Posting Date	Reference	Merchant Name or Transaction Description	New Purchases, Fees, Advances & Debits	Payments & Credits
5/05	5/08	495039873858-6	Cromwell Books	$32.89	
5/10	5/12	500003289887-6	Hasty Pudding	$65.00	
5/11	5/18	500003234788-9	World of Toys	$139.37	
	5/20		Payment, Thank You		$275.00
5/12	5/21	500035987983-0	Excelsior Cuts	$49.50	
5/19	5/21	500018398782-9	Dr. Carlton	$135.00	
5/19	5/21	500293879380-4	Weetwind Travel	$463.79	
5/19	5/22	500390029883-7	Cromwell Books	$19.84	
5/21	5/25	500039093784-1	World of Toys	$30.22	
5/21	5/25	500039851283-2	Dew Drop Inn	$69.47	

Previous Balance	New Purchases, Fees, Advances & Debits	Finance Charge (Due to Periodic Rate)	Payments & Credits	New Balance	Minimum Payment Due
$1161.23	$1004.88	$13.29	$275.00	$1904.40	$95.22

Monthly Periodic Rate	Corresponding Annual Percentage Rate	Balances to Which Applicable	FOR OFFICE USE ONLY
1.5%	18.00%	PURCHASES, ADVANCES, FINANCE CHARGES & FEES	

Student Page 97

The SUPER CHARGE CARD
D046025

Account Number	Statement Date
8322-156-00938	July 1

Roscoe and Vivian Van Dusen
2331 Grapevine Lane
Grand View, MN 55551

The Super Charge Card
Happy Valley State Bank
1425 W. Woodrow Wilson Blvd.
Montrose, TX 78712

Please indicate any changes below:
Name _____
Address _____
City, State _____ Zip Code _____
Telephone _____

Please be sure **our** mailing address above appears in the window of the return envelope.

Detach payment stub here.

Please include your account number on your check and on any correspondence.

Return this stub with your check.

Amount paid _____

Transaction Date	Posting Date	Reference	Merchant Name or Transaction Description	New Purchases, Fees, Advances & Debits	Payments & Credits
6/02	6/05	506893862798-0	Dr. Carlton	$173.00	
6/02	6/05	509874812894-8	Cromwell Books	$35.75	
6/03	6/05	507389827369-8	Dandy Candy	$19.23	
6/03	6/08	506839103927-7	Taetee Time Reataurant	$36.24	
6/04	6/17	502938472830-9	Scripps Brothers	$46.93	
	6/18		Payment, Thank You		$150.00
6/17	6/24	509828394732-1	HoneyDo Bakery	$31.23	
6/21	6/24	508392738413-6	Alfoneo's	$164.27	
6/21	6/24	509293092899-9	Hammer and Nale Hardware	$51.25	
6/22	6/24	510003974983-4	The Winslow Bank	$295.43	

Previous Balance	New Purchases, Fees, Advances & Debits	Finance Charge (Due to Periodic Rate)	Payments & Credits	New Balance	Minimum Payment Due
$1904.40	$853.34	$26.32	$150.00	$2634.06	$131.70

Monthly Periodic Rate	Corresponding Annual Percentage Rate	Balances to Which Applicable	FOR OFFICE USE ONLY
1.5%	18.00%	PURCHASES, ADVANCES, FINANCE CHARGES & FEES	

Student Page 98

Exercise IV. Internet Shopping

Today, it is possible to purchase almost anything on the Internet. You can find everything from CDs to automobiles on a variety of web sites at a variety of prices. What are the advantages and disadvantages of buying a product on the Internet rather than at a local store?

In the following exercise, you are asked to search for a number of products on the Internet. For each product, list the web site on which you found the item and the listed price. We have started the list with some common items that might interest teenagers. Add as many other items to the list as you care to.

ITEM	WEB SITE	PRICE
RCA 20" color TV		
Hockey elbow pads		
Child's mountain bike		
Evita soundtrack CD		
The Amber Spyglass (hardcover)		

There now exist many web sites for on-line shopping. Brand names, retail outlets, and prices vary widely from site to site and tend to change frequently.

Chapter 7

MORTGAGE LOANS

Teaching Notes

As students work on Schedules A through E, they may begin to realize that even a fraction of a percentage point can make a substantial difference in how rapidly a mortgage loan will be paid off.

Students may discover the algorithm mentioned on page 103 of the Activity Text as they work through amortization schedules with a hand calculator. That algorithm is as follows:

1. Enter the loan balance in memory.

2. Press the read memory (MR) key to display the balance.

3. Multiply the balance by the interest rate.

4. Divide the result displayed by 12 (the number of months in a year). Do not press the equals sign (=).

5. Press the subtract sign (–). The result displayed is the *interest* portion of the monthly payment.

6. Enter the monthly payment and press the equals sign (=). The result displayed is the *principal* portion of the monthly payment. It will appear as a negative value.

7. Add the result displayed (the *principal* portion of the monthly payment) to memory (M+).

8. Press read memory (MR) to display the new balance at the end of the month.

You may want to mention to students that banks usually figure interest payments daily. Each month the bank multiplies the loan balance by the annual percentage rate and divides by 360 days. (Some banks use 365 days, but most banks use 360 and earn a little extra interest by forgetting five days in the year.) The bank then multiplies the daily interest by the number of days between monthly payments (usually 29, 30, or 31 days). These calculations produce results that are similar to, but not exactly equal to, those produced by the method used in this book.

For Schedules D and E, a monthly payment has been invented for students to use in their calculations. Without knowing what interest rates students will discover, there is no way of knowing whether the monthly payments will actually be sufficient to amortize the loans. If you have access to a book of mortgage payment tables, you can determine the actual monthly payment needed to amortize the loans. Your local title company will have copies of such books, which they may allow you to see.

Answers to Text Questions, page 101

1. A *mortgage loan* is a loan made by a financial institution to a person, couple, or group of people who want to buy a house, but who do not have enough financial resources of their own. The financial institution agrees to pay the difference between the down payment on the house (generally 10–20% of the full price) and the full price of the house. The borrower agrees to pay some specific amount of money per month (or sometimes biweekly) that includes payment toward the principal and the interest. The house is collateral for the mortgage loan. If payments are not made as contracted, the financial institution may repossess the house.

2. (a) *Principal* refers to the total amount of money borrowed.

   (b) *Interest* is the amount of money paid each month to the financial institution at a specified rate, for the privilege of borrowing the principal.

   (c) *Points* represent a certain additional interest charge made on the mortgage loan. One point is equal to one percent of the loan amount.

(d) An *amortization schedule* is the program of payments set up over a certain number of months, showing how the money paid each month is divided between principal and interest, and how much remains to be paid on the principal.

(e) *Equity* is the dollar value of the house held by the buyer(s). Equity consists of the down payment plus all payments made to principal over the part of the amortization schedule that has been completed.

(f) *Escrow* is a fund or deposit set aside to cover specific costs associated with the sale, e.g., property taxes; also the period of time between which an agreement has been reached between seller and buyer and the time when the sale is actually completed (when "papers are passed").

3. The answer to this question depends upon the length of time the mortgage has been in effect. At the beginning of the mortgage period, most of the payment goes to interest. Toward the end of the loan period, most of the payment goes to principal. Students can appreciate this trend by examining a complete amortization schedule of the kind available at many web sites on the Internet.

4. *Closing costs* include all those expenses that arise as a result of the actual transfer of ownership of a house from one individual (or group of individuals) to another. Included are such charges as an appraisal fee; mortgage-recording fees; legal fees; and title insurance.

5. Students will have no *a priori* way of knowing the answer to this question. It will serve as an interesting springboard for a discussion on the topic. Students might be asked to calculate interest payments under both kinds of mortgages for short periods of time. Or you might invite an official from a local financial institution to discuss this point with students. Pros and cons of both types of mortgages are treated on a number of web sites.

6. Fixed-rate mortgages offer the advantage of a constant interest rate over the lifetime of the loan. The primary disadvantage is that prevailing interest rates may drop during the lifetime of the loan but the borrower must continue to pay the higher fixed rate. The interest rate on adjustable-rate mortgages changes constantly so that the borrower is always paying close attention to the going rate for loans. They may be at a disadvantage if interest rates climb throughout the lifetime of the loan, not an issue with a fixed-rate mortgage.

Comments on Internet Activities, page 101

1. Students should have no trouble finding web sites that list home-mortgage rates nationally and by region. A good general source of such web sites is the Look Smart consolidator on the Internet at http://www.looksmart.com/. Go to the category Work & Money and then follow through the subdivisions Personal Finance, Real Estate, Mortgages, and Quotes & Rates. The results students find will, of course, vary.

2. A number of web sites provide mortgage calculators that students can use knowing only the loan amount, interest rate, and loan term (years). Two fine examples of such calculators can be found at the following sites:

   (a) Loan Amortization Calculator: http://www.mortgage101.com/, followed by clicking on Amortization Tables.

   (b) http://www.interestratesonline.com/, followed by pressing the Mortgage Calculators key.

Student Page 102

Exercise I. Calculation of Mortgage Amount

Use the form below to calculate the total amount of money that must be borrowed on the purchase of a house that costs $186,700.

A.	Total cost of house	$186,700.00
B.	Down payment (20% of total cost of house)	$ 37,340.00
C.	Amount to be financed by mortgage (A–B)	$ 149,360.00
D.	Closing costs	

	1. Points (3% of mortgage)	$ 4,480.80
	2. Credit report	$75.00
	3. Application fee	$125.00
	4. Escrow fees	$872.00
	5. Notary fee	$25.00
	6. Title insurance	$915.00

E.	Total closing costs	$ 6,492.80
F.	Total loan amount (C + E)	$ 155,852.80

Use the form below to calculate the total amount of money that must be borrowed on the purchase of a house that costs $236,500.

A.	Total cost of house	$236,500.00
B.	Down payment (20% of total cost of house)	$ 47,300.00
C.	Amount to be financed by mortgage (A–B)	$ 189,200.00
D.	Closing costs	

	1. Points (2.5% of mortgage)	$ 4,730.00
	2. Credit report	$75.00
	3. Application fee	$75.00
	4. Escrow fees	$1,387.75
	5. Notary fee	$25.00
	6. Title insurance	$1,185.75

E.	Total closing costs	$ 7,478.50
F.	Total loan amount (C + E)	$ 196,678.50

Student Page 103

Exercise II. Amortization Schedules: Fixed- and Adjustable-Rate Mortgages

The following schedules let you compare the payments on a mortgage loan for one year at different rates of interest. Fill in the blanks in each schedule. Then you can compare the three mortgage arrangements. To figure the amounts for each month, follow these steps:

1. Amount toward interest = (previous balance x interest rate) ÷ 12

2. Amount toward principal = total payment – amount toward interest

3. New principal balance = previous balance – amount toward principal

Hint: Try to find an *algorithm* (system or method) by which you can carry out these calculations in the simplest possible way with the fewest steps.

Schedule A: 11½% Fixed Interest

Payment Number	Total Payment	Toward Interest	Toward Principal	Principal Balance
				$155,852.80
1	$1,532.00	$ 1,493.59	$ 38.41	$155,814.39
2	$1,532.00	$ 1,493.22	$ 38.78	$ 155,775.61
3	$1,532.00	$ 1,492.85	$ 39.15	$ 155,736.46
4	$1,532.00	$ 1,492.47	$ 39.53	$ 155,696.93
5	$1,532.00	$ 1,492.10	$ 39.90	$ 155,657.03
6	$1,532.00	$ 1,491.71	$ 40.29	$ 155,616.74
7	$1,532.00	$ 1,491.33	$ 40.67	$ 155,576.07
8	$1,532.00	$ 1,490.94	$ 41.06	$ 155,535.01
9	$1,532.00	$ 1,490.54	$ 41.46	$ 155,493.55
10	$1,532.00	$ 1,490.15	$ 41.85	$ 155,451.70
11	$1,532.00	$ 1,489.75	$ 42.25	$ 155,409.45
12	$1,532.00	$ 1,489.34	$ 42.66	$ 155,366.79
Totals		$ 17,897.99	$ 486.01	

Student Page 104

Schedule B: 11¼% Fixed Interest

Payment Number	Total Payment	Toward Interest	Toward Principal	Principal Balance
				$155,852.80
1	$1,532.00	$ 1,461.12	$ 70.88	$155,781.92
2	$1,532.00	$ 1,460.46	$ 71.54	$ 155,710.38
3	$1,532.00	$ 1,459.78	$ 72.22	$ 155,638.16
4	$1,532.00	$ 1,459.11	$ 72.89	$ 155,565.27
5	$1,532.00	$ 1,458.42	$ 73.58	$ 155,491.69
6	$1,532.00	$ 1,457.73	$ 74.27	$ 155,417.42
7	$1,532.00	$ 1,457.04	$ 74.96	$ 155,342.46
8	$1,532.00	$ 1,456.34	$ 75.66	$ 155,266.80
9	$1,532.00	$ 1,455.63	$ 76.37	$ 155,190.43
10	$1,532.00	$ 1,454.91	$ 77.09	$ 155,113.34
11	$1,532.00	$ 1,454.19	$ 77.81	$ 155,035.53
12	$1,532.00	$ 1,453.46	$ 78.54	$ 154,956.99
Totals		$ 17,488.19	$ 895.81	

Schedule C: Adjustable-Rate Mortgage

Payment Number	Total Payment	Toward Interest	Toward Principal	Principal Balance
				$155,852.80
1 @ 7.75%	$1,081.62	$ 1,006.55	$ 75.07	$155,777.73
2	$1,081.62	$ 1,006.06	$ 75.56	$ 155,702.17
3	$1,081.62	$ 1,005.58	$ 76.04	$ 155,626.13
4	$1,081.62	$ 1,005.09	$ 76.53	$ 155,549.60
5	$1,081.62	$ 1,004.59	$ 77.03	$ 155,472.57
6	$1,081.62	$ 1,004.09	$ 77.53	$ 155,395.04
7 @ 10.25%	$1,383.16	$ 1,327.33	$ 55.83	$ 155,339.21
8	$1,383.16	$ 1,326.86	$ 56.30	$ 155,282.91
9	$1,383.16	$ 1,326.37	$ 56.79	$ 155,226.12
10	$1,383.16	$ 1,325.89	$ 57.27	$ 155,168.85
11	$1,383.16	$ 1,325.40	$ 57.76	$ 155,111.09
12	$1,383.16	$ 1,324.91	$ 58.25	$ 155,052.84
Totals		$ 13,988.72	$ 799.96	

Student Page 105

> ### Answers to Schedules D and E will vary, depending on results of students' research.

For the next two schedules, call a local bank and find out the mortgage rates currently available for FRM and ARM loans. Then use those rates to complete these two schedules, as you did in Schedules A–C.

Schedule D: Current FRM Interest _____%

Payment Number	Total Payment	Toward Interest	Toward Principal	Principal Balance
				$155,852.80
1	$	$	$	$
2	$	$	$	$
3	$	$	$	$
4	$	$	$	$
5	$	$	$	$
6	$	$	$	$
7	$	$	$	$
8	$	$	$	$
9	$	$	$	$
10	$	$	$	$
11	$	$	$	$
12	$	$	$	$
Totals		$	$	

Schedule E: Current ARM Interest _____%

Payment Number	Total Payment	Toward Interest	Toward Principal	Principal Balance
				$155,852.80
1	$	$	$	$
2	$	$	$	$
3	$	$	$	$
4	$	$	$	$
5	$	$	$	$
6	$	$	$	$
7	$	$	$	$
8	$	$	$	$
9	$	$	$	$
10	$	$	$	$
11	$	$	$	$
12	$	$	$	$
Totals		$	$	

Chapter 8

HOME REPAIRS AND IMPROVEMENTS

Teaching Notes

Visiting lumberyards, hardware stores, and other sources of supplies provides a good introduction to shopping for materials. However, such visits may not be possible. You may prefer to have one student or a small group make the visits. Or you may send one student or a small group to one store, a second student or group to another, and so on. Students can then report to the whole class the results of their shopping expeditions.

The information needed for this chapter can also be obtained by searching the Internet. Depending on the size of your community, students may find prices at local stores. Or, they may have to use prices provided by on-line shopping sites or in nearby large cities. The Internet provides a good way for students to compare prices for supplies from various sources.

Comparison shopping, either in person or on the Internet, can provide students with an idea about the variety of prices and product quality available in the marketplace. This exercise provides the teacher with an opportunity to talk about the relative value

of products that are inexpensive but of poor quality compared to those that are more expensive but of higher quality.

The number of square feet covered by a can of paint (items 1c and 2i on page 110) differs for various paints. You can either have students use the number they find for any given kind of paint, or you can give them a number to use. The solution given on our answer pages assumes a coverage of 400 square feet per one-gallon can.

Stores often provide a discount for large quantities of materials purchased. For example, a discount of 10% for the purchase of more than ten cans of paint might be available. Have students inquire about such discounts and see if they will qualify for them in this exercise.

The prices of materials and supplies obviously varies widely in different parts of the country and over time. The prices given in our answers are intended purely as examples. The prices you and your students find in your own community may be quite different from those suggested here.

Student Page 108

Front View

Scale: 1 cm = 5'

Rear View

Scale: 1 cm = 5'

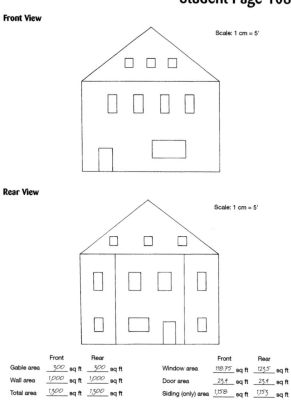

	Front	Rear		Front	Rear
Gable area	300 sq ft	300 sq ft	Window area	118.75 sq ft	123.5 sq ft
Wall area	1,000 sq ft	1,000 sq ft	Door area	23.4 sq ft	23.4 sq ft
Total area	1,300 sq ft	1,300 sq ft	Siding (only) area	1,158 sq ft	1,153 sq ft

Student Page 109

Window Detail (not to scale)

Although dimensions of various windows differ, they all have the same general design shown here. For this exercise, assume that the average outside dimensions are 36" wide and 56" high.

Area of window	2,016 sq in
Area of glass	1,581 sq in
Trim area to paint	435 sq in
(Total divided by 144) =	3 sq ft

Side View (both sides)

(Roof)

Scale: 1 cm = 5'

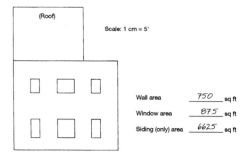

Wall area	750 sq ft
Window area	87.5 sq ft
Siding (only) area	662.5 sq ft

Example

Student Page 110

Worksheet for Cost of Painting House

1. Painting the Siding
   - (a) Number of square feet to be covered — 3636 sq ft
   - (b) Square feet of coverage needed for two coats (multiply (a) by 1.6) — 5817.6 sq ft
   - (c) Number of square feet covered by one gallon of house paint (see instructions on a paint can) — 400 sq ft
   - (d) Number of one-gallon cans needed for two coats — 15 cans
   - (e) Price per one-gallon can of paint — $ 18.95
   - (f) **Total cost of house paint** — $ 284.25

2. Painting Doors and Window Trim
   - (a) Area to be painted on one door — 23.4 sq ft
   - (b) Total number of doors — 2
   - (c) Total area of door trim to be covered — 46.8 sq ft
   - (d) Area to be painted on one window — 3 sq ft
   - (e) Total number of windows — 30
   - (f) Total area of window trim to be covered — 90 sq ft
   - (g) Total square feet of doors and window trim to be covered — 136.8 sq ft
   - (h) Square feet of coverage needed for two coats (multiply g. by 1.6) — 218.9 sq ft
   - (i) Number of square feet covered by one gallon of trim paint (see instructions on a paint can) — 400 sq ft
   - (j) Number of one-gallon cans needed for two coats — 1 cans
   - (k) Price per one-gallon can of trim paint — $ 19.95
   - (l) **Total cost of trim paint** — $ 19.95

   Total cost of all paint (add 1(f) and 2(l)) — $ 304.20

Example

Student Page 111

3. Other materials

   List and give the costs of other materials needed in painting. Some examples are given to start you out.

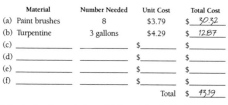

	Material	Number Needed	Unit Cost	Total Cost
(a)	Paint brushes	8	$3.79	$ 30.32
(b)	Turpentine	3 gallons	$4.29	$ 12.87
(c)			$	$
(d)			$	$
(e)			$	$
(f)			$	$
			Total	$ 43.19

4. Estimate of total cost of *all* materials for painting the Van Dusen house.

   Total $ 347.39

Example

Student Page 114

A. Expenses for Porch (Plan A)

1. Wood (sizes)	Amount Needed		Cost per Linear Foot	Total Cost
(a) 2" x 4"	39	lin ft	$ 1.19	$ 46.41
(b) 2" x 8'	140	lin ft	$ 1.49	$ 208.60
(c) 1" x 4" decking (36 pcs, 8 ft long)	288	lin ft	$ 0.59	$ 169.92
(d)		lin ft	$	$

2. Nails (sizes)	Amount Needed		Cost per Pound	Total Cost
(a)	50	lb	$ 0.90	$ 45.00
(b)		lb	$	$
(c)		lb	$	$
(d)		lb	$	$

3. Paint or Stain (colors)	Area to Paint	Number of Gallons	Price per Gallon	Total Cost
(a) creosote (for wood that touches the ground)		1	$ 5.98	$ 5.98
(b)			$	$

4. Other Materials (name)	Amount Needed	Price per Unit	Total Cost
(a)		$	$
(b)		$	$
(c)		$	$
(d)		$	$

Total of all costs $ 475.91

Answers will vary

Student Page 116

C. Expenses for Porch (Plan C)

1. Wood (sizes)	Amount Needed		Cost per Linear Foot	Total Cost
(a)		lin ft	$	$
(b)		lin ft	$	$
(c)		lin ft	$	$
(d)		lin ft	$	$

2. Nails (sizes)	Amount Needed		Cost per Pound	Total Cost
(a)		lb	$	$
(b)		lb	$	$
(c)		lb	$	$
(d)		lb	$	$

3. Paint or Stain (colors)	Area to Paint	Number of Gallons	Price per Gallon	Total Cost
(a)			$	$
(b)			$	$

4. Other Materials (name)	Amount Needed	Price per Unit	Total Cost
(a)		$	$
(b)		$	$
(c)		$	$
(d)		$	$

Total of all costs $

Answers will vary.

Student Page 115

B. Expenses for Porch (Plan B)

You might want to design a porch using different woods and paints, but keeping the same general shape and design. Compare the costs of this option with the first design. Use this form and the one on page 116 for your estimates.

1. Wood (sizes)	Amount Needed		Cost per Linear Foot	Total Cost
(a)		lin ft	$	$
(b)		lin ft	$	$
(c)		lin ft	$	$
(d)		lin ft	$	$

2. Nails (sizes)	Amount Needed		Cost per Pound	Total Cost
(a)		lb	$	$
(b)		lb	$	$
(c)		lb	$	$
(d)		lb	$	$

3. Paint or Stain (colors)	Area to Paint	Number of Gallons	Price per Gallon	Total Cost
(a)			$	$
(b)			$	$

4. Other Materials (name)	Amount Needed	Price per Unit	Total Cost
(a)		$	$
(b)		$	$
(c)		$	$
(d)		$	$

Total of all costs $

Example

Student Page 117

D. Patio Options

In order to make a fair comparison with the porch option, you will have to make the patio a certain size. When you have decided the size it will be, use those dimensions in the forms below.

Amount of land to be covered by patio: 240 sq ft

1. Solid cement patio

Materials Needed	Unit Cost	Total Cost
(a) 2 bags of cement	$ 7.95	$ 15.90
(b) 1 ton of sand	$ 41.90	$ 41.90
(c)	$	$
(d)	$	$
(e)	$	$
Total Cost		$ 57.80

2. Flagstone patio

Size of one flagstone	18" x 24"
Number of flagstones needed	80
Cost per flagstone	$ 0.75
Total cost of flagstones	$ 60.00

Other materials needed and cost of each:

sand to fill	$ 9.75
	$
	$
	$
	$
Total of all costs	$ 69.75

[Example]

Student Page 118

3. Brick patio

Size of one brick   _8' x 4'_

Number of bricks needed   _1080_

Cost per brick $ _0.315_

Total cost of bricks $ _340.20_

Other materials needed and cost of each:

_____sand to fill_____ $ ___9.75___

_____ $ _____

_____ $ _____

_____ $ _____

_____ $ _____

Total of all costs $ _349.95_

Chapter 9

UTILITY AND FUEL BILLS

Teaching Notes

Answers to Text Questions, page 120

1. Power companies generate electricity by a variety of means, such as coal- or oil-powered plants, nuclear power plants, hydro-electric plants, and thermal plants. The cost of electricity may vary according to the method used. Economic conditions vary across the country; power costs vary, as well. Demand for electrical power varies by region, by season, and by year. These differences result in differing prices for electrical power. Students may suggest other answers to this question.

2. A monopoly is a company that has the exclusive ability in an area to sell a particular product. There are federal laws designed to prevent the development of such market-dominant companies. Utility companies have long been allowed to exist as monopolies, however, because of the supposed benefits they provide to customers by offering a single source for electricity, telephone service, or some other service. In exchange for their monopoly status, utility companies are generally regulated by state or local governments.

3. Some of the extra services offered by many telephone companies are call waiting, call transfer, caller identification, special equipment, unlisted numbers, yellow-pages listing, extended-area calling, and telephone calling card.

   Federal and state charges may include a service-provider number charge, federal-access charge, federal universal-service fund, federal excise tax, state or local 911 charge, state universal-service charge, and residential-service protection fund.

4. Budget plans for heating-oil customers allow families to spread out the cost of cold-weather heating bills over the whole year, reducing the otherwise heavy financial impact during the winter. On the other hand, a severe winter may mean that budget payments do not adequately cover heating costs, and the customer may still have to pay more during the winter. Some people find it difficult to think about paying for heating expenses in the middle of a hot summer and prefer to wait until the cold weather arrives before thinking about such bills.

5. A service contract is an agreement between a company and an individual or family in which the company agrees to provide certain specific services at some flat rate. Those services often include an annual or biannual inspection and cleaning service and a certain number of free or low-cost emergency visits. The cost of the service contract is usually paid once a year and would not appear on bills other than that one month, although some companies bill service contracts biannually or on some other schedule.

6. A *therm* is a unit used to measure the quantity of heat provided by some fuel. It is now the standard of measurement for heat used by consumers in the United States. A therm is equal to 100,000 Btu (British thermal units), the unit of measure formerly used for this purpose.

7. The gas used to heat homes is normally natural gas, primarily methane with small amounts of other gaseous hydrocarbons.

Comments on Internet Activities, page 121

1. Students can get to web sites dealing with energy conservation in the home by beginning with any number of keywords or combinations of keywords, such as *energy conservation, home heating systems,* and *saving energy.* Two examples of helpful sites are the following:

   Alliance to Save Energy
   http://www.ase.org/

   Learning about Saving Energy
   (U.S. Department of Energy site)
   http://www.eren.doe.gov/erec/
   factsheets/savenrgy.html

2. Many energy companies provide their rate structure on their web sites. An especially good example is the web site for the American Electric Power company at http://www.aep.com/. State, regional, and local power companies have useful sites as well, for example, see the web site for Tucson Electrical Power at http://www.tucsonelectric.com/.

3. Student answers will vary.

4. Diesel oil, kerosene, gasoline, and various grades of heating oil are all produced in the fractional distillation of crude oil. They are all used as fuels, but for different, specific applications. Gasoline, for example, is used primarily for motor vehicles; kerosene is used for space heaters and in some kinds of generators. The numbers used in connection with home heating oils (such as *#6 heating oil*) denote the degree to which the fuel has been refined. Generally speaking, #2 heating oil is preferred for home-heating purposes. More information about heating oil can be found on the following web sites:

   Fuel Oil Facts
   http://www.fueloil.com/consumer/
   oilheat.html

   Oilheat.com
   http://oilheat.com/handbook/

Student Page 122

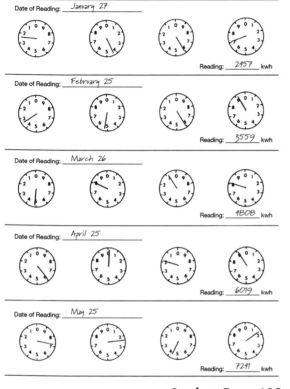

Date of Reading: _January 27_

Reading: _2457_ kwh

Date of Reading: _February 25_

Reading: _3559_ kwh

Date of Reading: _March 26_

Reading: _4808_ kwh

Date of Reading: _April 25_

Reading: _6019_ kwh

Date of Reading: _May 25_

Reading: _7241_ kwh

Student Page 123

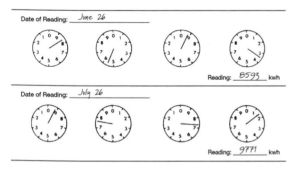

Date of Reading: _June 26_

Reading: _8593_ kwh

Date of Reading: _July 26_

Reading: _9771_ kwh

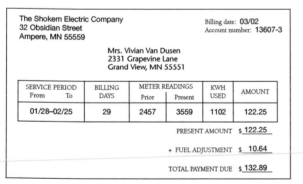

The Shokem Electric Company
32 Obsidian Street
Ampere, MN 55559

Billing date: **03/02**
Account number: **13607-3**

Mrs. Vivian Van Dusen
2331 Grapevine Lane
Grand View, MN 55551

SERVICE PERIOD From To	BILLING DAYS	METER READINGS Prior	METER READINGS Present	KWH USED	AMOUNT
01/28–02/25	29	2457	3559	1102	122.25

PRESENT AMOUNT $ **122.25**

+ FUEL ADJUSTMENT $ **10.64**

TOTAL PAYMENT DUE $ **132.89**

Student Page 124

The Shokem Electric Company
32 Obsidian Street
Ampere, MN 55559

Billing date: **04/02**
Account number: **13607-3**

Mrs. Vivian Van Dusen
2331 Grapevine Lane
Grand View, MN 55551

SERVICE PERIOD From To	BILLING DAYS	METER READINGS Prior	METER READINGS Present	KWH USED	AMOUNT
02/26–03/26	29	3559	4808	1249	139.93

PRESENT AMOUNT $ **139.93**

+ FUEL ADJUSTMENT $ **9.84**

TOTAL PAYMENT DUE $ **149.77**

The Shokem Electric Company
32 Obsidian Street
Ampere, MN 55559

Billing date: **05/02**
Account number: **13607-3**

Mrs. Vivian Van Dusen
2331 Grapevine Lane
Grand View, MN 55551

SERVICE PERIOD From To	BILLING DAYS	METER READINGS Prior	METER READINGS Present	KWH USED	AMOUNT
03/27–04/25	30	4808	6019	1211	135.36

PRESENT AMOUNT $ **135.36**

+ FUEL ADJUSTMENT $ **11.82**

TOTAL PAYMENT DUE $ **147.18**

Student Pa[ge]

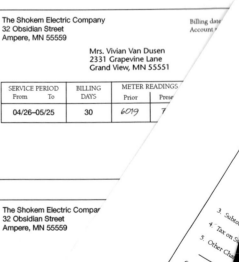

The Shokem Electric Company
32 Obsidian Street
Ampere, MN 55559

Billing date
Account

Mrs. Vivian Van Dusen
2331 Grapevine Lane
Grand View, MN 55551

SERVICE PERIOD From To	BILLING DAYS	METER READINGS Prior	Prese
04/26–05/25	30	6019	7

The Shokem Electric Compan
32 Obsidian Street
Ampere, MN 55559

SERVICE PERIO From		
05/26–		

2.

1/

1/

3. Subtota

4. Tax on S

5. Other Cha

6. **Total Charge**

Please detach and re
Account Number: 2
Enter Amount Paid:
Roscoe Van Dusen
2331 Grapevine Lane
Grand View, MN 55551

Student Page 126

The Shokem Electric Company
32 Obsidian Street
Ampere, MN 55559

Billing date: 08/02
Account number: 13607-3

Mrs. Vivian Van Dusen
2331 Grapevine Lane
Grand View, MN 55551

SERVICE PERIOD From To	BILLING DAYS	METER READINGS Prior	METER READINGS Present	KWH USED	AMOUNT
06/27–07/26	30	8593	9771	1178	131.39

PRESENT AMOUNT $ _131.39_

+ FUEL ADJUSTMENT $ _5.42_

TOTAL PAYMENT DUE $ _136.81_

Exercise II. Telephone Bills

Telephone companies have a variety of ways of charging for telephone calls. In the rate system chosen by the Van Dusens, the first 25 message units in each month are included in the Basic Monthly Rate, for which they pay $12.60. The cost of each message unit beyond the basic 25 is billed at 7.2¢. Use this information to complete the Van Dusens' telephone bills for the months shown on the following statements.

Student Page 127

Quad States Communications
P.O. Box 3999
Ringwell, MN 55549

Our Telephone Number: 1-800-555-5676

Customer Information:
Roscoe Van Dusen
2331 Grapevine Lane
Grand View, MN 55551
(218) 555-2698

Account Number: 218-555-2698-NJ989-07

Statement Date: 2/1/02

1. Quad States Communications Charges
   (a) Basic Monthly Rate, for period ending _1/27_ $ _12.60_
   (b) No. of Message Units (@ 7.2¢) _35_ $ _2.52_

Long Distance Charges $ _17.07_

1/4	Amarillo, TX	$3.95
1/9	New Thorville, AZ	$3.58
19	Enid, OK	$2.69
7	Chicago, IL	$3.05
7	Westport, TX	$3.80

of Charges $ _32.19_

btotal (@ 8%) $ _2.58_

ges $ _0.00_

Due (Please pay this amount in full.) $ _34.77_

- -
urn this portion with your payment.
8-555-2698-NJ989-07   Date Due: _2/15/02_

$_____

Quad States Communications
P.O. Box 3999
Ringwell, MN 55549

Student Page 128

Quad States Communications
P.O. Box 3999
Ringwell, MN 55549

Our Telephone Number: 1-800-555-5676

Customer Information:
Roscoe Van Dusen
2331 Grapevine Lane
Grand View, MN 55551
(218) 555-2698

Account Number: 218-555-2698-NJ989-07

Statement Date: 3/1/02

1. Quad States Communications Charges
   (a) Basic Monthly Rate, for period ending _2/16_ $ _12.60_
   (b) No. of Message Units (@ 7.2¢) _40_ $ _2.88_

2. Long Distance Charges $ _20.73_

2/2	Enid, OK	$4.20
2/6	Chicago, IL	$4.05
2/15	Dallas, TX	$4.20
2/15	Dallas, TX	$4.85
2/15	Westport, TX	$3.43

3. Subtotal of Charges $ _36.21_

4. Tax on Subtotal (@ 8%) $ _2.90_

5. Other Charges $ _0.00_

6. **Total Charges Due** (Please pay this amount in full.) $ _39.11_

- -
Please detach and return this portion with your payment.
Account Number: 218-555-2698-NJ989-07   Date Due: _3/15/02_

Enter Amount Paid: $_____

Roscoe Van Dusen
2331 Grapevine Lane
Grand View, MN 55551

Quad States Communications
P.O. Box 3999
Ringwell, MN 55549

Student Page 129

Quad States Communications
P.O. Box 3999
Ringwell, MN 55549

Our Telephone Number: 1-800-555-5676

Customer Information:
Roscoe Van Dusen
2331 Grapevine Lane
Grand View, MN 55551
(218) 555-2698

Account Number: 218-555-2698-NJ989-07

Statement Date: 4/1/02

1. Quad States Communications Charges
   (a) Basic Monthly Rate, for period ending _3/16_ $ _12.60_
   (b) No. of Message Units (@ 7.2¢) _27_ $ _1.94_

2. Long Distance Charges $ _19.79_

2/21	Amarillo, TX	$4.75
2/25	Enid, OK	$5.76
3/7	Enid, OK	$2.60
3/7	New York, NY	$2.85
3/9	Seattle, WA	$2.60
3/9	Dallas, TX	$1.23

3. Subtotal of Charges $ _34.33_

4. Tax on Subtotal (@ 8%) $ _2.75_

5. Other Charges $ _0.00_

6. **Total Charges Due** (Please pay this amount in full.) $ _37.08_

- -
Please detach and return this portion with your payment.
Account Number: 218-555-2698-NJ989-07   Date Due: _4/15/02_

Enter Amount Paid: $_____

Roscoe Van Dusen
2331 Grapevine Lane
Grand View, MN 55551

Quad States Communications
P.O. Box 3999
Ringwell, MN 55549

Student Page 130

Quad States Communications
P.O. Box 3999
Ringwell, MN 55549

Our Telephone Number: 1-800-555-5676

Customer Information:

Roscoe Van Dusen
2331 Grapevine Lane
Grand View, MN 55551
(218) 555-2698

Account Number: 218-555-2698-NJ989-07

Statement Date: 5/1/02

1. Quad States Communications Charges
   (a) Basic Monthly Rate, for period ending __4/16__ $ __12.60__
   (b) No. of Message Units (@ 7.2¢) __103__ $ __7.42__

2. Long Distance Charges $ __22.65__

4/2	Enid, OK	$11.83
4/7	Westport, TX	$6.19
4/10	St. Paul, MN	$0.37
4/11	St. Paul, MN	$1.69
4/11	Kansas City, KS	$2.57

3. Subtotal of Charges $ __42.67__

4. Tax on Subtotal (@ 8%) $ __3.41__

5. Other Charges and Credits $ __59.70__

   Installation of new line and jack $49.75
   Call-forwarding installation $9.95

6. **Total Charges Due** (Please pay this amount in full.) $ __105.78__

- -

Please detach and return this portion with your payment.

Account Number: 218-555-2698-NJ989-07

Date Due: __5/15/02__

Enter Amount Paid: $ _____

Roscoe Van Dusen
2331 Grapevine Lane
Grand View, MN 55551

Quad States Communications
P.O. Box 3999
Ringwell, MN 55549

Student Page 131

Exercise III. Heating Bills: Oil

The forms below can be used to calculate the cost of heating the Van Dusen home with oil. The items included in each bill are the amount of oil delivered and its price on the date of delivery (Del), any payments made by the customer (Payment), and any charge for the customer's service contract (Serv Cont).

The forms show that the Van Dusens are *budget* customers. That is, they pay a certain fixed amount each month throughout the year. What is the budget payment that they make each month?

Use the charges and payments shown on these bills to calculate the cost of heating the Van Dusen home with oil over a period of four months.

Hotshot Oil Company
P.O. Box 2289
St. John, MN 55547

M/M Roscoe Van Dusen
2331 Grapevine Lane
Grand View, MN 55551

(218) 555-4477

Account 449-38-22-6

Date	Reference	Code	Quantity	Unit Price	Charges	Credits	Balance
1/13	Del	01	189.40	$1.2370	$ *234.29*		$ *234.29*
	Prev Bal				$145.60		$ *379.89*
1/23	Payment	20				$100.00	$ *279.89*
1/25	Serv Cont	06			$42.00		$ *321.89*

Code: 01 – Fuel Oil   02 – Kerosene   03 – No. 5 Oil   04 – No. 6 Oil
05 – Gasoline   06 – Service Contract   07 – Other Fuel
08 – Equipment   09 – Repairs   10 – Adjustments   20 – Payment

Account Number: 449-38-22-6

Cash Payment Due: $ *321.89*
OR
Budget Payment Due: $ 100.00
Past Due or Prepaid: $ –0–
Current Budget
Payment Due: $ 100.00

Student Page 132

Hotshot Oil Company
P.O. Box 2289
St. John, MN 55547

M/M Roscoe Van Dusen
2331 Grapevine Lane
Grand View, MN 55551

(218) 555-4477

Account 449-38-22-6

Date	Reference	Code	Quantity	Unit Price	Charges	Credits	Balance
2/16	Del	01	182.50	$1.2490	$ *227.94*		$ *227.94*
	Prev Bal				$ *321.89*		$ *549.83*
2/20	Payment	20				$100.00	$ *449.83*

Code: 01 – Fuel Oil   02 – Kerosene   03 – No. 5 Oil   04 – No. 6 Oil
05 – Gasoline   06 – Service Contract   07 – Other Fuel
08 – Equipment   09 – Repairs   10 – Adjustments   20 – Payment

Account Number: 449-38-22-6

Cash Payment Due: $ *449.83*
OR
Budget Payment Due: $ 100.00
Past Due or Prepaid: $ –0–
Current Budget
Payment Due: $ 100.00

Student Page 133

Hotshot Oil Company
P.O. Box 2289
St. John, MN 55547

M/M Roscoe Van Dusen
2331 Grapevine Lane
Grand View, MN 55551

(218) 555-4477

Account 449-38-22-6

Date	Reference	Code	Quantity	Unit Price	Charges	Credits	Balance
	Prev Bal				$ *449.83*		$ *449.83*
3/9	Payment	20				$100.00	$ *349.83*
3/10	Del	01	205.60	$1.2390	$ *254.74*		$ *604.57*

Code: 01 – Fuel Oil   02 – Kerosene   03 – No. 5 Oil   04 – No. 6 Oil
05 – Gasoline   06 – Service Contract   07 – Other Fuel
08 – Equipment   09 – Repairs   10 – Adjustments   20 – Payment

Account Number: 449-38-22-6

Cash Payment Due: $ *604.57*
OR
Budget Payment Due: $ 100.00
Past Due or Prepaid: $ –0–
Current Budget
Payment Due: $ 100.00

Student Page 134

Hotshot Oil Company P.O. Box 2289 St. John, MN 55547				M/M Roscoe Van Dusen 2331 Grapevine Lane Grand View, MN 55551			
(218) 555-4477				Account 449-38-22-6			

Date	Reference	Code	Quantity	Unit Price	Charges	Credits	Balance
4/8	Payment	20				$100.00	$ −100.00
	Prev Bal				$ 604.57		$ 504.57
4/29	Del	01	102.22	$1.2300	$ 125.73		$ 630.30

Code:	01 – Fuel Oil	02 – Kerosene	03 – No. 5 Oil	04 – No. 6 Oil
	05 – Gasoline	06 – Service Contract	07 – Other Fuel	
	08 – Equipment	09 – Repairs	10 – Adjustments	20 – Payment

Account Number: 449-38-22-6	Cash Payment Due: OR	$ 630.30
	Budget Payment Due:	$ 100.00
	Past Due or Prepaid:	$ –0–
	Current Budget Payment Due:	$ 100.00

Exercise IV. Heating Bills: Gas

The gas used by people to heat their homes is measured with meters like those shown on the next page. Someone from the gas company reads the meter once a month. From these readings, the company can calculate how much gas has been used in the house. Then they can calculate how much the bill for that gas should be. Use the meter readings shown to calculate the gas bills for the months of January through April. The cost of the gas is shown in the chart below.

Zephyr Gas Company Service Classification #1: Residential	
First 5 therms or less	$10.50
6–35 therms	$0.3985 per therm
More than 35 therms	$0.8442 per therm
Minimum charge, exclusive of adjustments	$10.50

Student Page 135

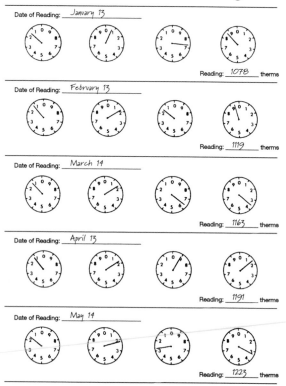

Date of Reading: January 13

Reading: 1078 therms

Date of Reading: February 13

Reading: 1119 therms

Date of Reading: March 14

Reading: 1163 therms

Date of Reading: April 13

Reading: 1191 therms

Date of Reading: May 14

Reading: 1223 therms

Student Page 136

Zephyr Gas Company 2145 Crestline Drive St. Vincent, MN 55507		Service Period: 1/14–2/13
Mr. and Mrs. Roscoe Van Dusen 2331 Grapevine Lane Grand View, MN 55551		Customer Number: 7891-BGH-4883

METER READINGS		QUANTITY USED	TOTAL CHARGES (from below)
Previous	Present		
1078	1119	41	$27.53

CHARGES

5	therms (Minimum Charge)	$10.50
30	therms @ 0.3985	$ 11.96
6	therms @ 0.8442	$ 5.07
	Total Charges	$ 27.53

Please mail your check to the address above.

Zephyr Gas Company 2145 Crestline Drive St. Vincent, MN 55507		Service Period: 2/14–3/14
Mr. and Mrs. Roscoe Van Dusen 2331 Grapevine Lane Grand View, MN 55551		Customer Number: 7891-BGH-4883

METER READINGS		QUANTITY USED	TOTAL CHARGES (from below)
Previous	Present		
1119	1163	44	$30.06

CHARGES

5	therms (Minimum Charge)	$10.50
30	therms @ 0.3985	$ 11.96
9	therms @ 0.8442	$ 7.60
	Total Charges	$ 30.06

Please mail your check to the address above.

Student Page 137

Zephyr Gas Company 2145 Crestline Drive St. Vincent, MN 55507		Service Period: 3/15–4/13
Mr. and Mrs. Roscoe Van Dusen 2331 Grapevine Lane Grand View, MN 55551		Customer Number: 7891-BGH-4883

METER READINGS		QUANTITY USED	TOTAL CHARGES (from below)
Previous	Present		
1163	1191	28	$19.67

CHARGES

5	therms (Minimum Charge)	$10.50
23	therms @ 0.3985	$ 9.17
0	therms @ 0.8442	$ –0–
	Total Charges	$ 19.67

Please mail your check to the address above.

Zephyr Gas Company 2145 Crestline Drive St. Vincent, MN 55507		Service Period: 4/14–5/14
Mr. and Mrs. Roscoe Van Dusen 2331 Grapevine Lane Grand View, MN 55551		Customer Number: 7891-BGH-4883

METER READINGS		QUANTITY USED	TOTAL CHARGES (from below)
Previous	Present		
1191	1223	32	$21.26

CHARGES

5	therms (Minimum Charge)	$10.50
27	therms @ 0.3985	$ 10.76
0	therms @ 0.8442	$ –0–
	Total Charges	$ 21.26

Please mail your check to the address above.

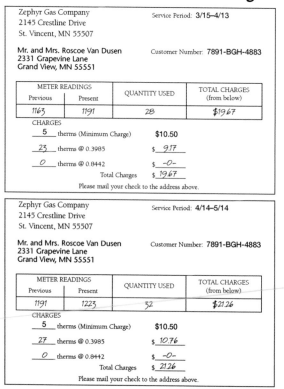

Student Page 139

| 3 | 6 | 1 | 1 |

Reading Date _5/6_

| 3 | 6 | 2 | 7 |

Reading Date _7/6_

| 3 | 6 | 4 | 4 |

Reading Date _9/7_

Grand View Water Department
Number One Civic Plaza
Grand View, MN 55551

Mr. and Mrs. Roscoe Van Dusen
2331 Grapevine Lane
Grand View, MN 55551

Account: 358-413-412

SERVICE PERIOD From To	BILLING DAYS	METER READINGS Prior	Present	WATER USED (ccf)
01/05–03/09	63	3578	3593	15

CHARGES

Water Used (@ $1.19 per ccf) $ _17.85_

Sewer Charges $ _17.89_

Extra Charges (see below) $ _0.00_

Total Due $ _35.74_

WATER CONSERVATION IS IMPORTANT! DO YOUR SHARE.

Your average daily water consumption this period:
(ccf used × 750, divided by no. of days)   _179_ gal

Your allotted daily consumption this period: _192_ gal

Amount your consumption exceeded allotment (if any):   _____ gal

Percent over allotted amount: _____ %

Penalty charges (if any): $_____

Student Page 140

Grand View Water Department
Number One Civic Plaza
Grand View, MN 55551

Mr. and Mrs. Roscoe Van Dusen
2331 Grapevine Lane
Grand View, MN 55551

Account: 358-413-412

SERVICE PERIOD From To	BILLING DAYS	METER READINGS Prior	Present	WATER USED (ccf)
03/10–05/06	58	3593	3611	18

CHARGES

Water Used (@ $1.19 per ccf) $ _21.42_

Sewer Charges $ _17.89_

Extra Charges (see below) $ _30.18_

Total Due $ _69.49_

WATER CONSERVATION IS IMPORTANT! DO YOUR SHARE.

Your average daily water consumption this period:
(ccf used × 750, divided by no. of days)   _233_ gal

Your allotted daily consumption this period: _192_ gal

Amount your consumption exceeded daily allotment (if any):   _41_ gal

Percent over daily allotted amount: _21.35_ %

Penalty charges (if any): $ _30.18_

Student Page 141

Grand View Water Department
Number One Civic Plaza
Grand View, MN 55551

Mr. and Mrs. Roscoe Van Dusen
2331 Grapevine Lane
Grand View, MN 55551

Account: 358-413-412

SERVICE PERIOD From To	BILLING DAYS	METER READINGS Prior	Present	WATER USED (ccf)
05/07–07/06	61	3611	3627	16

CHARGES

Water Used (@ $1.19 per ccf) $ _19.04_

Sewer Charges $ _17.89_

Extra Charges (see below) $ _.49_

Total Due $ _37.42_

WATER CONSERVATION IS IMPORTANT! DO YOUR SHARE.

Your average daily water consumption this period:
(ccf used × 750, divided by no. of days)   _197_ gal

Your allotted daily consumption this period: _192_ gal

Amount your consumption exceeded daily allotment (if any):   _5_ gal

Percent over daily allotted amount: _2.6_ %

Penalty charges (if any): $ _.49_

Student Page 142

Grand View Water Department
Number One Civic Plaza
Grand View, MN 55551

Mr. and Mrs. Roscoe Van Dusen
2331 Grapevine Lane
Grand View, MN 55551

Account: 358-413-412

SERVICE PERIOD From To	BILLING DAYS	METER READINGS Prior	Present	WATER USED (ccf)
07/07–09/07	63	3627	3644	17

CHARGES

Water Used (@ $1.19 per ccf) $ _20.23_

Sewer Charges $ _17.89_

Extra Charges (see below) $ _2.00_

Total Due $ _40.12_

WATER CONSERVATION IS IMPORTANT! DO YOUR SHARE.

Your average daily water consumption this period:
(ccf used × 750, divided by no. of days)   _202_ gal

Your allotted daily consumption this period: _192_ gal

Amount your consumption exceeded daily allotment (if any):   _10_ gal

Percent over daily allotted amount: _5.21_ %

Penalty charges (if any): $ _2.00_

Chapter 10

AUTOMOBILE EXPENSES

Teaching Notes

This chapter is somewhat different from other chapters in *Math in Everyday Life* in that so many variables are involved in students' calculations. Students will find, for example, that car prices differ not only from maker to maker, model to model, and year to year, but also from dealer to dealer and area to area. The same can be said for gasoline prices, tire prices, the cost of repairs, and the cost of insurance.

For this reason, the exercises in this chapter may have a different emphasis from those in other chapters. Students can still be expected to carry out a number of mathematical calculations involved in the purchase or leasing of a car. However, they can also be encouraged to think about the host of decisions in buying or leasing a car and the guidelines that may be available in making wise choices.

The Text Questions and Internet Activities may be of even greater value than in other chapters. They can act as springboards for discussions about the many decisions to be made in car purchases and leasing.

One value in carrying out the mathematical exercises in this chapter may be to encourage students to think about the costs involved in car ownership or leasing; the ways in which such costs can be paid; and the ways in which they can be reduced.

You have many options for carrying out the exercises in this chapter. You may ask individual students or groups of students to visit automobile showrooms, gasoline stations, repair shops, tire stores, and other facilities to obtain the prices they need to complete these activities. Or you can do that research yourself or invent reasonable numbers that can then be given to students to use in the exercises.

You may choose to have students use different values within an exercise. For example, one might want to use a higher grade of gasoline (at a higher price) in a newer car than in an older car. One might also make different kinds of insurance decisions for cars of different ages. Following this path can make calculations somewhat more difficult, but they do let students think about real-life financial decisions and may, therefore, have additional learning value.

Answers to Text Questions, pages 144–145

1. Among some of the questions one might ask a salesperson: (a) What kind of warranty is available on the car? (b) What financing plans can be arranged? (c) What gasoline mileage will the car give? (d) What kinds of extras are available, and what is the price of each? (e) What kind of trade-in value might be expected on the car? (f) What kinds of services are available from the dealer?

2. A *warranty* is a guarantee by the car manufacturer or the company selling the car that any mechanical problems that develop in the car during a certain period of time will be corrected by the seller at no cost to the buyer. The conditions covered by a warranty vary from time to time and from company to company, depending on how eager a company is to sell its cars. In slow economic conditions, warranty conditions are likely to be better than when cars are selling well.

3. Unbiased consumer organizations test cars and publish information about the way they perform. One of the oldest and best known is Consumers Union, which publishes test results in its magazine *Consumer Reports* (also available on-line for a fee at http://www.consumerreports.org/). Various automotive magazines also do extensive testing of cars each year. Two of the best known are *New Car Test Drive* (also at http://www.newcartestdrive.com/) and *Car and Driver* (also at http://www.caranddriver.com/). In addition, the National Highway Traffic Safety Administration conducts regular tests of a large variety of automobiles. Their web site is at http://www.nhtsa.dot.gov/cars/.

4. (a) *Bodily injury liability* covers the cost of medical bills up to a certain limit for serious physical injuries resulting from an accident. Payments are made on the policyholder's behalf if someone is hurt by or in the policyholder's car or in another car.

(b) *Automobile medical payments* coverage is similar to *bodily injury liability,* but for less serious physical injuries.

(c) *Property damage liability* is insurance that protects a policyholder if his or her car is involved in an accident that causes damage to someone else's public or private property (car, house, fence, utility pole, etc.).

(d) *Comprehensive* insurance pays for the damage to a car as a result of fire, theft, or vandalism. The insurance company usually pays the estimated amount of the damage, less the deductible. (For *deductible*, see question 6.)

(e) *Collision* insurance pays for damage to a car if that car is involved in an accident and the person wants the insurance company to pay for repairs. Such occasions may arise if an accident is determined to be the policyholder's fault or if another driver involved in the accident does not have adequate insurance coverage. The insurance company normally pays the estimated damage, less the deductible.

(f) *Uninsured motorist* coverage protects a person and passengers if the policyholder's car is struck by a car whose driver does not carry any (or enough) liability insurance.

5. Some of the factors that should be considered in deciding on the type of coverage include the following: the number of passengers who usually travel in the car; the age, value, and weight of the car; the speed at which the car is normally driven; the number of miles the car is driven each year; the chance of a serious accident; and the ability of the car owner to absorb the costs of an accident. If the car owner has financed the purchase of a vehicle with a bank, credit union, or auto credit agency, he or she may be required to have certain types and amounts of coverage.

6. The *deductible* on an insurance policy is the amount that policyholder must pay before the insurance company covers any other insurance costs. For example, a person who has a *$100 deductible* on any part of a policy agrees to pay $100 toward the cost of any accident in which he or she is involved. Policies without deductible clauses or with low deductibles are more expensive than others. Without such deductible clauses, insurance companies would find themselves paying for every scratch and dent made in a car.

7. Some of the factors that account for wide discrepancy in insurance premiums include the following: the risk of driving in a particular area (urban versus rural, for example); competition among insurers (many insurers versus a few, for example); differences in age, sex, driving experience, and traffic violations; differences in age and condition of vehicles; and financial decisions made by insurers themselves.

8. *No-fault* insurance means that no attempt is made to find out who is at fault in an accident. Damage to a car (and sometimes the injuries resulting from an accident) are paid directly by an insurance company. The details of specific no-fault insurance plans vary from state to state. Some states provide information about their programs on their web sites. A good general source of information is the National Association of Insurance Commissioners, whose web site (`http://www.naic.org`) contains an extensive bibliography of articles on no-fault insurance.

9. Radial and glass- or steel-belted tires are built to last a long time. Of the two, radials probably give the most control to a car. On the other hand, they are also the most expensive. Glass- and steel-belted tires are good for heavy cars in which sharp turning is not a problem. One should take into consideration the age of the car when buying tires because

very good tires may last longer than the car for which they are purchased.

Comments on Internet Activities, page 146

1. Students can locate many web sites dedicated to the purchase of vehicles on-line. The usual procedure is for the customer to select, in order, the make of car, model, and options desired. The customer then enters personal information, such as name, address, and telephone number. The web site provider then notifies a car showroom in the customer's geographic area about the request. A salesperson then follows up with a telephone call or e-mail to the prospective customer.

   Typical of the many sites dedicated to this service are the following:

   > A Car Online
   > http://www.acaronline.com/
   >
   > Auto Buying Service
   > http://www.autoweb.com/
   >
   > Car Buying Tips
   > http://www.carbuyingtips.com/car2.htm

   Web sites of this kind frequently offer related services, such as car insurance, financing, and a variety of warranties.

2. There are many approaches to the purchase of automobile insurance on the Internet. Many insurance companies and agents maintain their own web pages, on which they advertise the policies and rates available. Some web sites promise to search a variety of companies and find the best price available on car insurance for the customer. Some companies do business almost exclusively on the Internet.

   Purchasing auto insurance on the Internet is similar to the face-to-face process. The prospective customer provides personal data (such as name, address, age, gender, and smoking/non-smoking status) and information about the vehicle to be insured (such as make, model, and year). A price quote is then prepared by the insurance company or the information is submitted to companies that wish to offer bids on the policy.

   One advantage of buying insurance on the Internet is that it is often possible to find a low rate that would take a great deal of research to find otherwise. A disadvantage is that service from a company located on the Internet may be more difficult to obtain than service from a local agent.

3. An excellent resource for advice on purchasing car insurance is the Look Smart section of the Internet. This section contains directions to web sites that offer on-line car insurance as well as web sites that provide information on wise practices in the purchase of car insurance. The address for this site is http://www.looksmart.com/. The hierarchy to follow is Work & Money ➡ Personal Finance ➡ Insurance ➡ Auto Insurance ➡ Quotes & Rates.

4. Fuel-efficiency information on the Internet is available at http://www.fueleconomy.gov/.

Student Page 146

Internet Activities

1. It is now possible to shop for new and used cars on the Internet. Find out how to get price quotes and purchase a car on the Internet.

2. Is it possible to purchase car insurance on the Internet? If so, what are the advantages and disadvantages of doing so?

3. Are you able to locate advice on the Internet about wise practices in the purchase of automobile insurance? If so, what are some of the points you learned in your search?

4. Each year, the U.S. Environmental Protection Agency (EPA) publishes a list of vehicles for sale in the United States and the gas mileage to be expected of each type of vehicle. Locate this information on the Internet and list the three most fuel-efficient and the three least fuel-efficient vehicles as determined by the EPA.

Note: Before beginning the exercises in this chapter, please note that the figures used here were made up for the purpose of this exercise. They are not intended to represent real differences among various makes of car, which, in any case, are likely to change over time.

Exercise I. Purchasing and Financing a Car

Craig Van Dusen has information about buying four cars he likes. If he makes a down payment of $500, he can get a 36-month auto loan to finance the rest of the cost. For each car, calculate the amount to be financed and the annual cost of payments on the loan. (The monthly payment includes interest on the loan.)

Car	Cost	Down Payment	Amount to Be Financed	Monthly Payment	Annual Total
Honda (new)	$19,989	$500	$ *19,489* at 3½% APR	$585.72	$*7,028.64*
Ford (2-yr-old)	$ 8,375	$500	$ *7,875* at 7½% APR	$244.96	$*2,939.52*
Chevrolet (3-yr-old)	$ 6,999	$500	$ *6,499* at 7¼% APR	$201.41	$*2,416.92*
Plymouth (5-yr-old)	$ 3,500	$500	$ *3,000* at 8¼% APR	$ 94.36	$*1,132.32*

Student Page 147

Example

Exercise II. Maintaining a Car

A. Annual Operating Costs

1. Gasoline Costs

Car	Miles per Gallon	Miles per Week	Driven per Year	Gallons of Gasoline Used	Price per Gallon*	Total Cost of Gasoline
Honda	35.2	125	*6500*	*185*	$1.69	$*312.65*
Ford	23.9	125	*6500*	*272*	$1.69	$*459.68*
Chevrolet	20.7	125	*6500*	*314*	$1.69	$*530.66*
Plymouth	19.8	125	*6500*	*328*	$1.69	$*554.32*

*Check at local station for current price.

2. Oil-Change Costs

Car	Miles per Oil Change	Number of Oil Changes per Year	Cost per Oil Change*	Total Cost of Oil Changes
Honda	3,000	*3*	$*19.95*	$*59.85*
Ford	5,000	*2*	$*19.95*	$*39.90*
Chevrolet	4,500	*2*	$*19.95*	$*39.90*
Plymouth	3,000	*3*	$*19.95*	$*59.85*

*Check at local station for current price.

3. Tires

Kind of Tire	Expected Lifetime (in miles)**	Cost per Tire**	Cost per Mile (per tire)
A	*18,000 miles*	$*50.50*	$*0.0028*
B	*30,000 miles*	$*62.75*	$*0.0021*
C	*45,000 miles*	$*109.90*	$*0.0024*

**See question 9 on page 145.

4. Repairs

Car	Average Expected Cost of Repairs per Mile	Miles per Week	Miles per Year	Total Cost
Honda	3.1¢	125	*6500*	$*201.50*
Ford	4.5¢	125	*6500*	$*292.50*
Chevrolet	7.2¢	125	*6500*	$*468.00*
Plymouth	9.7¢	125	*6500*	$*630.50*

Student Page 150

Example

Driver Variations

The rate tables on the preceding pages show the premiums for a driver over the age of twenty-five who drives less than ten miles to work one way. However, these premiums change for younger drivers and for those who drive more than ten miles one way to work. To adjust the premiums for those categories of drivers, multiply the premiums by the percentage shown in the appropriate category. You can write your answers on the Actual Premium line at the bottom of each car-insurance worksheet.

	Driver Under Age 25	Driver Over Age 25
Driving to and from work less than 10 miles one way	135%	Basic Premium
Driving to and from work more than 10 miles one way	175%	128%

Worksheet for Car 1: Honda (new)

Type of Coverage	Limits of Coverage		Premium
A. Bodily Injury Liability			
Each person $ *25,000* / Each accident $ *50,000*			$ *182.76*
B. Property Damage Liability	Each accident $ *25,000*		$ *96.42*
C. Automobile Medical Payments	Each person $ *25,000*		$ *23.00*
D. Comprehensive	Deductible $ *100*		$ *174.00*
E. Collision	Deductible $ *250*		$ *337.60*
F. Towing and Labor			$ *25.00*
G. Uninsured Motorist Protection			
Each person $ *25,000* / Each accident $ *50,000*			$ *28.00*
		Total Premium	$ *866.78*
Driver Variation Factor: × *175* %		Actual Premium	$ *1,516.87*

Student Page 151

Examples

Worksheet for Car 2: Ford (2-yr-old)

Type of Coverage	Limits of Coverage		Premium
A. Bodily Injury Liability			
Each person $ *25,000* / Each accident $ *50,000*			$ *182.76*
B. Property Damage Liability	Each accident $ *25,000*		$ *96.42*
C. Automobile Medical Payments	Each person $ *25,000*		$ *23.00*
D. Comprehensive	Deductible $ *100*		$ *142.00*
E. Collision	Deductible $ *250*		$ *288.00*
F. Towing and Labor			$ *25.00*
G. Uninsured Motorist Protection			
Each person $ *25,000* / Each accident $ *50,000*			$ *28.00*
		Total Premium	$ *785.18*
Driver Variation Factor: × *175* %		Actual Premium	$ *1,374.07*

Worksheet for Car 3: Chevrolet (3-yr-old)

Type of Coverage	Limits of Coverage		Premium
A. Bodily Injury Liability			
Each person $ *25,000* / Each accident $ *50,000*			$ *182.76*
B. Property Damage Liability	Each accident $ *25,000*		$ *96.42*
C. Automobile Medical Payments	Each person $ *25,000*		$ *23.00*
D. Comprehensive	Deductible $ *100*		$ *78.00*
E. Collision	Deductible $ *250*		$ *168.00*
F. Towing and Labor			$ *25.00*
G. Uninsured Motorist Protection			
Each person $ *25,000* / Each accident $ *50,000*			$ *28.00*
		Total Premium	$ *601.18*
Driver Variation Factor: × *175* %		Actual Premium	$ *1,052.07*

Example

Student Page 152

Worksheet for Car 4: Plymouth (5-yr-old)

Type of Coverage	Limits of Coverage		Premium
A. Bodily Injury Liability			
Each person $ _25,000_ / Each accident	$ _50,000_		$ _182.76_
B. Property Damage Liability	Each accident	$ _25,000_	$ _96.42_
C. Automobile Medical Payments	Each person	$ _25,000_	$ _23.00_
D. Comprehensive	Deductible	$ _100_	$ _41.00_
E. Collision	Deductible	$ _250_	$ _132.00_
F. Towing and Labor			$ _25.00_
G. Uninsured Motorist Protection			
Each person $ _25,000_ / Each accident	$ _50,000_		$ _28.00_
		Total Premium	$ _528.18_
Driver Variation Factor: × _175_ %		Actual Premium	$ _924.32_

Exercise III. Total Annual Car Expenses

Item	Car 1	Car 2	Car 3	Car 4
Purchase and finance (not including down payment)	$7,028.64	$2,939.52	$2,416.92	$1,132.32
Gasoline costs	$312.65	$459.68	$530.66	$554.32
Oil-change costs	$59.85	$39.90	$39.90	$59.85
Tire costs (4 tires for 6,500 miles per year)	$72.80	$72.80	$72.80	$72.80
Other repairs	$201.50	$292.50	$468.00	$630.50
Insurance	$1,576.87	$1,374.07	$1,052.07	$924.32
Totals	$9,192.31	$5,178.47	$4,580.35	$3,374.11

Student Page 153

Exercise IV. Leasing a Car

Many people today prefer to lease rather than purchase a car. Companies that need to have large fleets of cars for their employees are especially likely to lease rather than buy those cars.

The way a leasing program works is that an automobile dealer sells a car to a leasing company. The leasing company can be a subdivision of the dealer itself, a subdivision of a large car manufacturer, or an independent leasing company. The consumer then leases the car from the leasing company.

Car leases are usually written for 24, 36, or 48 months. The cost of the lease is determined by the value of the car during the time it is leased plus interest charges on the lease. That is, suppose a car costs $20,000 new and is leased for a 24-month period. At the end of that period, the car is worth $12,000. The cost of the lease, then, is $8,000 ($20,000 − $12,000) plus interest charges. The monthly charges are then the total cost of the lease plus the interest charges divided by 24.

The chart below shows four possible new-car leasing plans that Craig has investigated. Determine the annual cost to Craig for each of the four leasing plans.

Note: Company names (first column) are as follows:

EZL: E-Z-Lease
#1: #1 Lease
Go$: Go$ Leases
L'n'S: Lease 'n' Save

A Company Name	B Initial Cost of Car	C Value of Car at End of Lease	D Cost of Lease (B – C)	E Leasing Period	F Total Interest Due	G Monthly Cost	H Annual Cost
EZL	$20,000	$9,000	$11,000	48 months	$1,852.00	$267.75	$3,213.00
#1	$25,000	$13,500	$11,500	36 months	$1,971.38	$374.21	$4,490.52
Go$	$36,275	$21,950	$14,325	48 months	$3,314.79	$367.50	$4,410.00
L'n'S	$21,350	$10,375	$10,975	24 months	$1,097.28	$503.01	$6,036.12

Chapter 11

LIFE INSURANCE

Teaching Notes

The subject of insurance is among the most difficult of all consumer math issues. Insurance contracts are notoriously complex, detailed, and opaque. Some companies have made earnest efforts to write their policies in everyday language that anyone can read. But even these efforts are sometimes unsuccessful.

For this reason, we recommend that you invite a professional into the classroom if at all possible. Agents are likely to be more than happy to speak to your class about the intricacies of life insurance, as well as of health- and auto-insurance policies.

You might ask the speaker to discuss not only the basic principles of insurance, but also some of the differences among types of policies and among companies, as well as the reasons for these differences. Variations in laws and procedures from state to state may also be of some interest to students. This presentation might stimulate a discussion as to the reasons that insurance is such a complex topic.

In addition, there are now a number of web sites that attempt to explain the fundamental principles of life insurance. A good place to begin is the Look Smart collection of sites, at http://www.looksmart.com/. Follow the directory from Work & Money ➜ Personal Finance ➜ Insurance ➜ Life Insurance ➜ Advice & Guides.

Answers to Text Questions, pages 155–156

1. (a) *Ordinary* or *straight* life insurance is paid into over lifetime until (usually) the age of sixty-five, at which time the total cash value of the policy is redeemable.

    (b) *Term* insurance is paid into for a certain specific number of years (the term of the policy), usually for ten or twenty years. The policy covers the holder only during that period of time. At the end of the period, coverage is terminated and there is no cash value refunded.

    (c) An *annuity* policy is one that is paid for in one lump sum or in several installments. The issuing company pays a regular pension on the investments made by the holder.

    (d) In an *endowment* policy, premiums are paid for a definite number of years. At the end of that period, the face value of the policy is paid to the holder.

2. For both Craig and Vanessa, the best policy might be an ordinary life policy because the premiums are relatively low compared to the coverage. Mr. Van Dusen might want to take out endowment policies for each of the children so that at age twenty, each would have a certain sum of money to put toward college expenses, if they chose to go. Both he and his wife should also have a policy, either ordinary life or a term policy, to protect the family against loss of income in case of the death of either parent. Decisions on the type of policy and the amount of coverage for each family member involve a number of factors. Students should discuss some of these factors and the circumstances in which each kind of policy would be most appropriate.

3. (a) The *premium* is the amount of money paid by the holder at regular intervals to the insurance company. It is calculated on a fixed rate, and there are usually several payment options, such as once a year or once a month.

    (b) The *policyholder* is the person who is covered by the policy. He or she often, but not always, initiates the policy and pays the premiums.

    (c) The *beneficiary* is the person to whom the face value of the policy is paid upon the death of the policyholder.

(d) The *face value* of the policy is the amount that is paid to the beneficiary upon the death of the policyholder.

(e) The *cash surrender value* is the amount of money that a policyholder can borrow, using the policy as collateral. It is also the amount that the holder can receive if she or he decides to cash in the policy.

(f) Some insurance companies are *mutual* companies, in which all policyholders also hold some share in the company itself. In such companies, each policyholder may receive a *dividend* at the end of the year that reflects any excess profits made by the company above and beyond a "reasonable" profit. The dividend may be paid in one of three ways: (1) by deducting the dividend from the premium due; (2) by leaving the dividend with the company to earn additional insurance or interest; or (3) by paying the dividend to the holder in cash.

(g) *Life expectancy tables* are charts that estimate the number of years remaining in the life of a person at any particular age. The calculation of life expectancy tables is a highly developed process whose accuracy is of the greatest importance to an insurance company. Such tables allow companies to estimate premiums for people of various ages that will be reasonable and yet will guarantee that the company will make a profit on its policies.

4. Traditionally, life-insurance policies have not been regarded as a particularly good form of savings. Historically, insurance policies earned about 2% interest, compared to interest rates close to 5% in many commercial banks. Today banks pay about the same rate on their savings accounts that insurance companies do on their insurance policies.

Strictly on the basis of interest rates, then, bank savings accounts may no longer be preferable to insurance policies.

In addition, life insurance policies provide death benefits and sometimes other forms of protection not available from commercial bank savings accounts.

However, it is difficult to withdraw money from a life insurance savings account compared to a bank savings account.

There are many safe methods for saving money that bring better interest rates than those available with insurance policies. Such methods include certificates of deposit, specialized bank savings accounts, stocks and bonds, and mutual funds.

Comments on Internet Activities, page 156

1. An excellent collection of web sites dealing with life insurance issues can be found on Look Smart, at http://www.looksmart.com/. Follow the directory from Work & Money ➔ Personal Finance ➔ Insurance ➔ Life Insurance ➔ Advice & Guides.

2. Loan calculators can be found on a number of Internet web sites. They will not all provide the same results for the information given in the Activity Text. As an example, the results obtained from the Insurance Cost Estimator on the BudgetLife web site are $504 for Mr. Yoshimoro and $380 for Mrs. Yoshimoro.

3. Dozens of life-insurance companies maintain their own web sites. Students may well know the names of insurance companies in your own community with which they can begin a search of the Internet. It is possible to find the names of other companies on the Web. As an example, the web site Life Insurance Companies lists a number of individual companies through which they operate. Find them at http://www.life-insurance-companies.com/.

Student Page 157

Exercise I. Choosing a Form of Life Insurance

The Van Dusens' insurance agent is trying to convince Craig to buy a life insurance policy. The agent points out that the younger Craig is when he buys a policy, the less it will cost him. Craig has agreed to learn more about life insurance policies. He has made a chart like the one below. The table on the next page provides all the information you need to complete Craig's chart.

After completing the chart, decide which policy Craig should buy *or* whether he would be better off waiting a few years to buy life insurance. Compare the costs of policies if purchased at age 18; age 23; and age 28.

A. Costs of Various Insurance Plans

If purchased now:	Ordinary Life	20-Year Endowment	10-Year Term
Cost of a $ 5,000 policy	$75.30	$243.50	$29.90
$10,000 policy	$135.60	$472.00	$49.80
$25,000 policy	$326.50	$1,167.50	$112.00
$50,000 policy	$643.00	$2,325.00	$214.00

If purchased in 5 years:			
Cost of a $ 5,000 policy	$83.85	$244.95	$32.10
$10,000 policy	$152.70	$474.90	$54.20
$25,000 policy	$369.25	$1,174.75	$123.00
$50,000 policy	$728.50	$2,339.50	$236.00

If purchased in 10 years:			
Cost of a $ 5,000 policy	$94.30	$246.65	$34.30
$10,000 policy	$173.60	$478.30	$58.60
$25,000 policy	$421.50	$1,183.25	$134.00
$50,000 policy	$833.00	$2,356.50	$258.00

Student Page 159

Exercise II. Borrowing Money Against a Life Insurance Policy

Mrs. Van Dusen is currently holding a $250,000 insurance policy that she took out at the age of 30. She is now 43. If she should decide to borrow money against that policy, how much could she get? Do not take into consideration any dividends or interest that might also have accumulated.

How much would she be able to borrow if she waited until next year? Until the year after that? Until she was 60 years old?

Suppose Mrs. Van Dusen had taken out her policy at the age of 25. What would be the comparable loan values for each of the above situations? Use the Table of Loan Values to find the answers. Put your answers in the following table.

Age at Which Policy Was Taken Out	VALUE		
	Now	Next Year	At Age 60
25	$73,795.00	$79,087.50	$166,815.00
30	$57,737.50	$63,530.00	$162,605.00

Table of Loan Values

CASH OR LOAN VALUE FOR $1,000 OF FACE AMOUNT OF POLICY (TYPICAL RATES)

End of Year	Age 18	Age 19	Age 20	Age 21	Age 22	Age 23	Age 24	Age 25	Age 26	Age 27	Age 28	Age 29	Age 30	Age 31
3	$ .00	$ .99	$ 1.95	$ 2.94	$ 3.98	$ 5.06	$ 6.19	$ 7.36	$ 8.59	$ 9.86	$11.18	$12.59	$14.04	$15.59
4	14.15	15.49	16.88	18.33	19.83	21.40	23.04	24.74	26.52	28.37	30.31	32.35	34.49	36.74
5	28.40	30.23	32.06	33.96	35.94	38.00	40.15	42.39	44.72	47.17	49.72	52.41	55.22	58.19
6	43.05	45.22	47.49	49.84	52.30	54.86	57.52	60.29	63.20	66.23	69.40	72.73	76.24	79.92
7	57.86	60.46	63.16	65.98	68.91	71.97	75.14	78.46	81.93	85.56	89.36	93.34	97.53	101.92
8	72.90	75.93	79.09	82.37	85.78	89.33	93.03	96.90	100.94	105.16	109.58	114.22	119.09	124.22
9	88.20	91.66	95.26	99.00	102.89	106.95	111.18	115.60	120.21	125.03	130.07	135.36	140.93	146.78
10	103.73	107.63	111.67	115.88	120.26	124.82	129.58	134.55	139.73	145.15	150.82	156.78	163.04	169.62
11	119.50	123.83	128.33	133.01	137.88	142.95	148.24	153.76	159.52	165.54	171.84	178.46	185.41	192.72
12	135.52	140.28	145.23	150.38	155.75	161.33	167.15	173.22	179.55	186.18	193.12	200.40	208.05	216.10
13	151.77	156.97	162.38	168.01	173.86	179.96	186.30	192.93	199.85	207.08	214.65	222.60	230.95	239.75
14	168.26	173.91	179.78	185.88	192.22	198.83	205.71	212.89	220.39	228.23	236.44	245.06	254.12	263.66
15	184.99	191.09	197.41	203.99	210.82	217.94	225.36	233.10	241.18	249.63	258.48	267.78	277.55	287.84
16	201.96	208.50	215.28	222.23	229.66	237.30	245.25	253.55	262.22	271.28	280.78	290.75	301.23	312.78
17	219.16	226.15	233.39	240.92	248.75	256.90	265.39	274.24	283.50	293.18	303.32	313.97	325.18	336.99
18	236.60	244.03	251.73	259.74	268.06	276.73	285.75	295.18	305.02	315.31	326.11	337.45	349.38	361.98
19	254.27	262.14	270.31	278.79	287.61	296.80	306.36	316.35	326.78	337.70	349.15	361.18	373.86	387.24
20	272.17	280.49	289.11	298.08	307.39	317.10	327.21	337.76	348.78	360.33	372.44	385.18	398.60	412.78
AT AGE														
60	684.59	682.46	680.22	677.88	675.42	672.84	670.12	667.26	664.25	661.07	657.72	654.17	650.42	646.45
65	795.30	795.30	795.30	795.30	795.30	795.30	795.30	795.30	795.30	795.30	795.30	795.30	795.30	795.30

The values applying to this policy are those in the column headed with the age corresponding to the insuring age shown above, do not include the value of any divided accumulations or additions which may be standing to the credit of this policy, and are for policies having no indebtedness.

Chapter 12

INCOME TAX

The subject of income taxes is so complex that it may be helpful to have a professional visit the classroom to talk with students. You may be able to get a representative of the IRS or an accountant to explain the income-tax system and answer student questions. Remember, however, not to ask for help during the two or three months before April! Tax forms are available at http://www.irs.gov.

Most students should be able to complete forms 1040EZ and 1040A, although 1040 may be too much for middle and high school students. You may wish to assign one part of the form to one group of students, a second part of the form to another group, and so on. Or you may ask students to get help from parents, neighbors, or friends in completing Form 1040 using simplified financial records.

The exercises in this chapter can be made simple or complex by deciding on the number and type of financial transactions to be entered in the monthly summary sheets.

If your state has an income tax, you may want to have students complete a copy of the state form; or you may prefer to have them simply examine the form and compare it to federal tax forms.

Tax forms typically change from year to year as the U.S. Congress passes new tax laws. The general information requested does tend to stay the same, however, as more tax reforms involve arcane issues. The forms presented in the Activity Text are the most recent available at the time of publication. You may wish to have students obtain tax publications and tax forms from a local IRS office, the local library, or on-line. These forms can then be used in place of those in the Activity Text.

Answers to Text Questions, page 162

1. Generally speaking, the three 1040 forms differ from each other in their complexity. Individuals with complex financial transactions, whose income is high, and who have a variety of expenses, are required to use Form 1040. Those with less complex financial transactions can often use Form 1040A. Finally, individuals whose income and expenditures are relatively simple may be able to use Form 1040EZ.

Each year, the IRS provides an instruction page called Which Form Should I Use, similar to the one on Activity Text page 177 that provides the specific requirements for the use of each form. Students should refer to the most recent edition of this page to find out current requirements.

2.–3. Some of the additional forms and schedules that may be needed in completing a tax return are the following:

  A For itemizing deductions

  B For dividends, interest, and other stock distributions in excess of some given amount

  C For income from a personally owned business

  D For income from sale or exchange of capital assets

  E For income/loss from pensions, annuities, rents, royalties, etc.

  R For credit for the elderly or the disabled

  SE For reporting net earnings from self-employment

  3903 For reporting moving expenses

  4684 For reporting gains and losses resulting from casualties and thefts

4. A W-2 form is prepared by an individual's employer, reporting the gross wages for the past year and the amounts of federal, state, and city income taxes, and the amounts of Social Security and Medicare taxes that have been withheld from the employee's wages.

A taxpayer should use the W-2 form to find out exactly how much she or he was paid and how much tax was withheld during the year. W-2 forms must be included with the taxpayer's 1040, 1040A, or 1040EZ report form.

5. Because tax code changes from one year to the next, and because individual circumstances vary, the answer to this question is not always clear. Generally speaking, married couples are likely to pay less in taxes if they file jointly than if they file separately. A factor is whether both husband and wife have separate, independent sources of income. In case of doubt, it is a good idea to figure taxes both ways (separately and jointly) to see which results in the lower tax.

6. Only certain expenses can be deducted from income tax if they exceed a certain minimum amount, for example, medical and dental expenses, mortgage interest, certain other types of interest, and most kinds of taxes.

Comments on Internet Activities, page 162

1. One of the first places to look for help on tax questions is the Internal Revenue Service's web site at http://www.irs.gov. A number of individuals, professional groups, and businesses also maintain web sites that contain useful information on income taxes, for example:

> Tax Forms Home
> http://www.savewealth.com/taxforms/
>
> Income Tax Law: Free Legal Information
> http://law.freeadvice.com/tax_law/income_tax_law/

> Look Smart categories
> http://www.looksmart.com/
> Follow the path: Work & Money ➜ Personal Finance ➜ Taxes ➜ Tax Firms ➜ Services (by letter).

2. Students should easily locate the web site from which they can obtain information about income taxes in your own state. Many states make it possible to download at least some tax forms and publications.

3. The *flat tax* concept is one in which each taxpayer pays a certain specific percentage of his or her annual income. The flat tax idea has been in circulation for many years. Students should read articles on the Internet both in support of and in opposition to the flat tax. Examples of the former can be found on the Look Smart web site by beginning with the address http://www.looksmart.com/ and then following the path Work & Money ➜ Personal Finance ➜ Taxes ➜ Tax Reform ➜ Flat Tax. An article in opposition to the flat tax can be found at Flat Tax Fiasco: http://www.wordwiz72.com/flattax.html.

A second alternative to the federal income tax is a national retail sales tax, similar in concept to state sales taxes. Arguments for and against this concept can be found at Idea House of the National Center for Policy Analysis (http://www.public-policy.org/~ncpa/) and on the Look Smart web site at http://www.looksmart.com/, and then: Work & Money ➜ Personal Finance ➜ Taxes ➜ Tax Reform ➜ National Retail Tax.

Share Your Bright Ideas

We want to hear from you!

Your name_____Date_____

School name_____

School address_____

City _____State _____Zip_____Phone number (_____)_____

Grade level(s) taught_____Subject area(s) taught_____

Where did you purchase this publication?_____

In what month do you purchase a majority of your supplements?_____

What moneys were used to purchase this product?

___School supplemental budget   ___Federal/state funding   ___Personal

Please "grade" this Walch publication in the following areas:

Quality of service you received when purchasing	A	B	C	D
Ease of use	A	B	C	D
Quality of content	A	B	C	D
Page layout	A	B	C	D
Organization of material	A	B	C	D
Suitability for grade level	A	B	C	D
Instructional value	A	B	C	D

COMMENTS:_____

_____

What specific supplemental materials would help you meet your current—or future—instructional needs?

_____

Have you used other Walch publications? If so, which ones?_____

May we use your comments in upcoming communications?   ___Yes   ___No

Please **FAX** this completed form to **888-991-5755**, or mail it to

Customer Service, Walch Publishing, P. O. Box 658, Portland, ME 04104-0658

We will send you a **FREE GIFT** in appreciation of your feedback. **THANK YOU!**